DATE DUE

			PRINTED IN U.S.A.

INVASIVE SPECIES

WHAT EVERYONE NEEDS TO KNOW

INVASIVE SPECIES

WHAT EVERYONE NEEDS TO KNOW

DANIEL SIMBERLOFF

OXFORD
UNIVERSITY PRESS

OXFORD

UNIVERSITY PRESS

Oxford University Press is a department of the University of Oxford.
It furthers the University's objective of excellence in research, scholarship,
and education by publishing worldwide.

Oxford   New York
Auckland   Cape Town   Dar es Salaam   Hong Kong   Karachi
Kuala Lumpur   Madrid   Melbourne   Mexico City   Nairobi
New Delhi   Shanghai   Taipei Toronto

With offices in
Argentina   Austria   Brazil   Chile   Czech Republic   France   Greece
Guatemala   Hungary   Italy   Japan   Poland   Portugal   Singapore
South Korea   Switzerland   Thailand   Turkey   Ukraine   Vietnam

Oxford is a registered trademark of Oxford University Press
in the UK and certain other countries.

Published in the United States of America by
Oxford University Press
198 Madison Avenue, New York, NY 10016

Library of Congress Cataloging-in-Publication Data
Simberloff, Daniel.
Invasive species : what everyone needs to know / Daniel Simberloff.
pages cm
Summary: "This book studies the ecology of invasive species, examining
the effects that such invasions have on various types of ecosystems"—
Provided by publisher.
ISBN 978–0–19–992201–7 (hardback)—ISBN 978–0–19–992203–1 (pbk.)
1. Introduced organisms.   2. Evolution (Biology)   3. Plants—Evolution.
4. Conservation biology.   I. Title.
QH353.S56 2013
578.6'2—dc23

CONTENTS

FOREWORD

In the late 1970s, I became interested in biological invasions as a purely academic phenomenon. I was studying aspects of community ecology, particularly how different species fit together (or do not fit together!) to form biological communities. Introduced species seemed to me a useful subject, a test of various ideas about communities. After all, a biological invasion consists of a new species arriving in the midst of an existing community of native species. Will the new entry survive or not, and, if it does, how will it affect the natives? It seemed to me that answers to these questions would shed light on the forces structuring ecological communities. How similar can an introduced species be to an existing native species yet still persist? Conversely, will similarity to a native species aid an invasive species in displacing that species? How does the number of native species already present affect the likelihood that a new introduction will survive and spread? What happens when a totally new life form is introduced to a community (as when rats arrive on a formerly predator-free island)?

As I combed the literature on invasions seeking data that might help answer such questions, I found a problematic situation. Although many researchers had noted particular invasions and some had speculated on the reasons for them and the impacts they caused, relatively few had treated invasions in enough detail to help me answer the general questions I was asking. In fact, it seemed as if the majority of invasions were highly idiosyncratic, with the impacts resting on details that were very specific to the particular introduced species and to the biota of the particular region to which it was introduced. I wondered if perhaps there were no general answers. My own empirical research at the time, which included observation of both invasive and native insects attacking invasive and native plants, was too narrowly focused to suggest general patterns. Some of my observations were intriguing. For instance, Asian oaks widely planted in north Florida displayed remarkable resistance to native leaf-mining moths and beetles attacking them compared with native oak species. But how relevant were such findings to general patterns of biological invasions?

As I pondered my research results and delved into the literature on biological invasions in the mid-1980s, I participated in two projects almost simultaneously that greatly influenced my research career and definitively shifted my focus toward invasions. In 1982, the Scientific Committee on Problems of the Environment (SCOPE), an arm of the International Council of Scientific Unions, recognized a growing number of reports of environmental problems caused by biological invasions and an absence of any sort of synthetic, scientific overview of the phenomenon. In response, SCOPE organized a series of workshops in several nations to encourage study of the causes and consequences of biological invasions and to attempt to apply this research to solving the myriad problems invasions cause. These workshops engaged many of the world's leading ecologists, both those who had studied particular invasions intensively and those who were less familiar with the subject. I was fortunate enough to be invited to the North American workshop, held in California in 1984, and the final synthesis workshop in 1988 in Hawaii, which brought together participants from the regional and national workshops. These events featured many fascinating presentations on details of various invasions and highly stimulating discussions in both the daily sessions and informal conversations at dinner and into the night.

Coincidentally, in 1986 I was asked to join the national Board of Governors of the Nature Conservancy. I had long been engaged in conservation activities, both as a private citizen and as a scientist publishing papers on refuge design and causes of extinction. Among its many activities, the Nature Conservancy maintains an enormous network of refuges, and after joining the board I quickly learned from many of the land stewards who were managing those refuges that introduced species were a major problem—often the biggest problem—in a remarkable fraction of them. Invasive plants that overgrow the entire habitat were the most prevalent management challenge, but there were many others, including introduced predators that threaten rare animals, introduced herbivorous insects and mammals that imperil plants, and introduced plants that change the fire cycle to the disadvantage of native species. Juxtaposing this information from land stewards fighting invasions on the ground in the conservation trenches with the wealth of data and theory that scientists in the SCOPE project provided, I could not help but recognize that biological invasions were a subject of great practical urgency and limitless scientific opportunity. My research path was set.

The SCOPE project of the 1980s thus launched the science of modern invasion biology, bringing together ecologists, evolutionists, geneticists, earth scientists, and mathematical modelers in an explosion of research aimed at filling in the details of long-noted invasions as well as detecting impacts of new invasions and discerning previously unrecognized effects of older invasions. By the 1990s, land managers were intensively engaged in invasion biology, working with various technologies to eradicate invasive species or at least mitigate their harmful impacts. In the 2000s, economists and social scientists have increasingly interacted with invasion biologists to address the human side of invasion impacts and management of introduced species. However, it is not easy for the public and policymakers to understand the wealth of issues that biological invasions raise. Every week, newspapers and television shows feature reports of biological invasions, usually of local or regional interest: a new insect pest of regional forests or crops; an introduced fish that has made its way into local lakes; a new plant overrunning gardens or nature preserves; a new mosquito carrying a human pathogen. Occasionally, a particularly damaging invasion receives national coverage. However, media reports tend to sensationalize stories, presenting grim possible future scenarios but few details and often no context. Each of these reports is usually a one-off event, and none of them present the panorama of biological invasions and management options that people must understand if they are to form rational opinions about specific invasions.

In addition, many aspects of biological invasion management are controversial. Sometimes management costs are high, and, in some cases, the degree of invasion impact is debatable. Management procedures involving baits or herbicides can have nontarget impacts, or an eradication technique such as snaring wild boar or feral pigs can be inhumane. And, as with any large environmental problem (global warming and chemical pollution are examples), a small minority of scientists will argue that the problem is exaggerated or management actions unwarranted—or both. The popular media focus on controversy, so critics of invasion biology get a lot of press, again without the context that everyone needs to assess the criticisms intelligently.

The increasing number of damaging biological invasions, the rapid growth of a modern science to study them, the wealth of old and new management technologies (often unpublicized), and frequent controversies over policies and practice are all factors that combine to create a need for a comprehensive, nontechnical discussion of the full scope of biological invasions. This book aims to satisfy that need.

ACKNOWLEDGMENTS

My students and colleagues at the University of Tennessee have been a continuing source of information and insights on biological invasions. Every day I hear from them a new finding on one or more invaders or a new idea about how to study and understand them. In particular, my former students Martin Nuñez, Arijana Barun, Betsy Von Holle, Diego Vázquez, Tad Fukami, and Lara Souza all worked on biological invasions for their doctorates and taught me much about the systems they studied. They constantly bombarded me with challenging questions and readings of the invasion literature. My current students, Sara Kuebbing, Rafael Zenni, Noelia Barrios-Garcia, and Jessica Welch, have carried on this tradition and keep me from thinking I know too much about invasions. Sara and Rafael read an entire early draft of this manuscript and provided many suggestions that helped me improve it; for this service I am greatly indebted.

My wife, Mary Tebo, my daughter Ruth Simberloff, and my cousin Carol Bodian read and constructively criticized selected chapters. My daughter Tander Simberloff exhaustively edited an entire draft of the manuscript and kept hauling me down from the ivory tower—I will be forever appreciative.

Colleagues too numerous to list answered myriad questions on particular invasions discussed in this book, often providing substantial unpublished details. I am grateful for their assistance.

My editor at Oxford University Press, Hallie Stebbins, has been a constant source of encouragement and advice. The University of Tennessee granted me a year-long development leave (known elsewhere as a sabbatical), and a French hotbed of research on invasions, the Centre d' Écologie Fonctionnelle et Évolutive of the Centre National de la Recherche Scientifique in Montpellier, was a congenial host during this period. I wrote this book during this leave, and I thank both institutions for their assistance.

INVASIVE SPECIES

WHAT EVERYONE NEEDS TO KNOW

1

GENERAL INTRODUCTION

1.1 What is a biological invasion?

On a sunny day in May 2002, two men went fishing at a small, shallow pond (originally a construction pit) behind a strip mall in Crofton, Maryland, 20 miles from the Capitol in Washington, D.C. Almost immediately, a large, strange fish snapped up a cast lure. This was unlike anything either man had seen before, resembling an obese, muscular eel, with fins running almost the entire length of the top and bottom of the 18-inch body and, clamped around the lure, a set of teeth like something out of a science fiction movie. Imbued with the catch-and-release philosophy, the fishermen released the fish, noting that it also looked pregnant. But, intrigued by this scary oddity, they photographed it before throwing it back in the water, though they did not develop the roll of film until June. None of their fishing buddies recognized the fish in the snapshot, so they took it to an office of the Maryland Department of Natural Resources. Even the fisheries scientists there could not identify it, but finally, through a website of nonnative fish species operated by the U.S. Geological Survey, they determined it to be a snakehead. The authors of the website then identified it as the northern snakehead, native to the Yangtze River in China north to the China–Russia border. This fish is a ferocious, aggressive predator that eats almost any other fish and can grow to 3 feet long. Worse, juveniles of this species can breathe air, survive awhile out of water, and move short distances overland—and the small Maryland pond is near the Little Patuxent River.

A month later, a fisherman caught a larger northern snakehead in the same Crofton pond. Soon after, he caught several baby snakeheads. Because of the proximity to Washington, the arrival of this fearsome fish made regional and then national headlines. Two horror movies were even loosely based on this event. It later turned out that this "invasion" had begun when a man purchased two individuals at an Asian food market and subsequently released them. The pond was drained, a hundred snakeheads of various sizes were destroyed, and none have been seen near the site since. However, a small breeding population was later discovered in the Potomac River (these individuals were determined by genetic means not to be related to those in Crofton), and individuals have cropped up in several distant states. Most ominously, they have been found in drainage ditches near the White River in Arkansas, from which they would have easy access to the Arkansas and Mississippi Rivers.

The arrival and spread of the northern snakehead is a classic biological invasion. The triggering event occurs when individuals of a species not native to a region arrive with human assistance and establish an ongoing population. If the population then spreads in its new home, the phenomenon is called a *biological invasion* and the species is termed *invasive*, at least in this region. Another example is the red imported fire ant, which arrived from South America in the port of Mobile, Alabama, in the 1930s and gradually spread throughout the southern United States, so that today it extends north to Tennessee and North Carolina and west to Texas. In the 1990s it jumped to southern California and began spreading there as well.

Several aspects of this definition require explanation. When we say "arrive with human assistance," this need not mean deliberate assistance. Many nonnative plant species are deliberately imported for horticultural use, and some, such as kudzu and Japanese knotweed, become invasive. Similarly, some deliberately introduced sport fish, such as the South American butterfly peacock bass in Florida, have become invasive. However, many invaders are inadvertently introduced as hitchhikers on goods carried or shipped by humans. The fire ant arrived in Mobile hidden in cargo and probably caught a ride to California in trucked plant material or soil. Additionally, the term *invader* is often used to describe all introduced species, but this is not quite accurate. An introduced species is any species that arrives somewhere with human help (deliberate or accidental), including even those species that do not establish populations or that establish populations but do not spread widely from the point of arrival. Biologists have generally restricted the term invasive to cases in which the species is found well beyond the arrival area.

By contrast with biologists' usage, policymakers sometimes use *invasive species* to describe only introduced species with negative impacts. For instance, President Bill Clinton's Executive Order 13112 of 1999 defines invasive species to be an "alien" species whose introduction does or is likely to harm the environment, the economy, or human health—a rather different definition from that traditionally used by biologists. Invasive species are sometimes referred to simply as "pests," but, in fact, not all invaders cause significant damage, and not all pests are introduced. In this book, invasive species will be used in the standard biological sense—species that arrive with human assistance, establish populations, and spread.

The word *native* also needs some explanation. Biologists say a species is native to a region if it evolved there or if it evolved elsewhere but arrived in the region by its own means, usually thousands if not millions of years ago and without human assistance. Species present in a region but not native have been termed *alien, exotic, nonindigenous,* and *nonnative,* but the various nonbiological uses, sometimes pejorative, of the terms *alien* and *exotic* have led biologists to use nonindigenous and especially nonnative for species not native to a region. For the great majority of biological invasions discussed in popular media and scientific literature, there is no question whether the species is in fact nonnative or might perhaps be native; its status is known. However, there are exceptions—some populations are suspected of having been introduced, but the evidence is simply insufficient to be conclusive. The common periwinkle snail of the northern United States and Canadian Atlantic coast is one such example that will be detailed in Chapter 11.3.

Populations whose status as native or nonnative is uncertain are termed *cryptogenic*, and for some regions or habitats the percentage of species that are cryptogenic is quite large. For instance, the native geographic range of many marine species is not known, and for widespread marine species it is often impossible to determine if they reached their current locations on their own or were carried in ballast water or on the hulls of ships, perhaps centuries ago before records were kept of which species were where. For some species, populations on the edge of what is believed to be the native range of a species may in fact be introduced. Among plants in the western Mediterranean region, several are ancient human imports from adjacent areas. For instance, the chasteberry, native to the eastern Mediterranean, was brought to Mediterranean France by monks for use as an anaphrodisiac to reduce libido. Among fishes in the United States, a surprising number of such "peripheral" populations were actually established by humans for sport fishing.

In much of the world, widespread human transport of species is a relatively recent phenomenon. Invasion biologists often focus on invasions that have occurred in the last 500 years, since the European discovery of America initiated what is known as the Columbian Exchange—the widespread movement of animals, plants, humans, and culture between the Old World and New World. Of course, humans moved species around well before that. For instance, the Lapita people carried plants and animals (including the Pacific rat) from southeast Asia to many islands in Melanesia, Micronesia, and Polynesia in the Pacific beginning around 3,000 years ago. Within the Americas, native peoples probably moved some species around as well. It is quite possible that at least one of the species termed *native fire ants* in the United States is actually a pre-Columbian introduction. In Europe, so many species have been moved around over previous millennia that European botanists use different terms for plant species introduced after 1500 (*neophytes*) and those introduced before 1500 (*archeophytes*).

A few species do not fall neatly into the categories of native or nonnative. This is not because of lack of knowledge but rather due to peculiarities of their history. For instance, in Scotland, the capercaillie or wood grouse was native, disappeared in 1775, then was deliberately reintroduced from Scandinavia in the 1830s. Should it be classified as native? The mountain wisent in the Caucasus, a subspecies of the European bison, was entirely exterminated by 1927 but was reestablished after World War II with imported individuals descended from the last remaining wild herd of the Polish subspecies and from zoos. However, because of previous breeding of the zoo animals with American bison, all individuals in the present herd have at least a small fraction of their genes from this different species. This fact led some to argue that the species should be eliminated from the Caucasus on the grounds that it is not native—it was both introduced from elsewhere and hybridized. This argument did not prevail, but the mountain wisent does not cleanly fall into either the native or nonnative category. Still, cases such as these, though interesting, are a very small fraction of all populations in nature. Most species either are clearly native to a place or got there with direct or indirect human help and are therefore nonnative.

Once in a great while, a species makes a tremendous discontinuous leap on its own, establishing a population in a region far from its original range and then spreading in its new home. The most famous example is the cattle egret, an Old World species that flew across the Atlantic Ocean from Africa. As with many bird species, individuals of this species had occasionally been seen far from their normal homes. Cattle egrets seen sporadically in northern South America in the 19th century fall in this category. Such species are termed *vagrants* by birdwatchers and are usually assumed to have been blown off course by unusual wind currents. However, in the 1930s a population established there and spread to North America in 1941. The species is now found widely in the United States as well as in Canada and Mexico. Some people refer to the spread of the cattle egret as an invasion, though technically it would not qualify as such. Such a dramatic, discontinuous range expansion resembles a biological invasion in every way except that it occurred without human aid. In any event, sudden long jumps of this sort are extremely rare.

1.2 When did the study of biological invasions begin?

The great public and scientific concern with biological invasions is a recent phenomenon, though there were many antecedents. By the 17th century, Englishmen at home and in the colonies had noted the presence of some introduced species. Perhaps the first person to study them scientifically was Pehr Kalm, a Swedish Finn sent by Linnaeus (the founder of modern taxonomy) in the 18th century to acquire American plants for acclimatization in Sweden. In America, Kalm recognized at least 15 European plants and the honeybee as well as other Old World insects. Throughout the 19th century, several scientists made note of introduced species in various places, especially the Swiss botanist Augustin de Candolle and his son Alphonse de Candolle. Alphonse de Candolle and the English amateur botanist Hewett C. Watson were the first to produce classifications of species as native, introduced, and what we would now call cryptogenic. Also in the 19th century, Charles

Darwin and his fellow evolutionists Alfred Russel Wallace and Joseph Hooker all noted and recorded introduced species in their travels. Darwin was concerned with more than just matters of geographic distribution. He lamented that in some cases invaders had virtually replaced formerly dominant native plants. He even addressed the question, which became of great scientific interest a century later, of why some introduced species establish populations and spread while others remain restricted or die out.

In the early 20th century, two persons wrote masterful, systematic tabulations of introduced species: James Ritchie in Scotland (*The Influence of Man on Animal Life in Scotland*, 1920) and George M. Thomson in New Zealand (*The Naturalisation of Animals and Plants in New Zealand*, 1922). Neither of these works had a major influence on scientists at the time, although Thomson's book was rediscovered toward the end of the 20th century and used to deduce several historical dates and pathways of spread of introduced species. In 1957, the Swede Carl Lindroth published *The Faunal Connections between Europe and North America*, a work chiefly on insects, which discussed many European species found in North America, with details on their locations and history. It is noteworthy that each of these 20th-century books dealt only with one specific region and only one of them (Thomson's) with both animals and plants.

In 1958, the English ecologist Charles Elton published a book on invasions aimed at a popular audience, *The Ecology of Invasions by Animals and Plants*, based on a series of BBC radio shows he had produced on the subject. This book was the culmination of Elton's lifelong interest in invasions. During World War II, to aid the war effort, he had turned his entire Oxford Bureau of Animal Population toward research on four introduced pests of British agriculture: the house mouse, the Norway rat, the ship rat, and the rabbit. In his book, Elton discussed both geographic distributions of invaders and their ecological impacts. He dealt with plants, animals, and microbes in terrestrial, freshwater, and marine species on continents and islands worldwide. In short, his was the first comprehensive treatment of biological invasions. Elton's book is sometimes referred to as the book that started invasion biology as a scientific discipline, and Elton has been called the father of invasion biology.

Elton's book addressed many of the ecological issues that greatly concern scientists and the public today. However, though many scientists read it in the years following its appearance, it did not crystallize the formation of a discipline of invasion biology. Elton was a man ahead of his time. Various people continued to conduct research on particular invasive species, and the public became concerned with certain invasions—a particular insect like the gypsy moth in North America, a particular agricultural weed like cheatgrass in the American West, a particular predatory mammal like the small Indian mongoose in the Caribbean and Hawaii, a particular plant pathogen like the chestnut blight in eastern North America. However, not until the 1980s were scientific efforts mobilized to address the entire gamut of biological invasions as a general phenomenon. Over the years, it had become increasingly obvious that more and more problems were being caused by invasive species. In response, the Scientific Committee on Problems of the Environment (SCOPE), an arm of the International Council of Scientific Unions, developed an international project on impacts and management of biological invasions, led by American biologist Harold Mooney. The SCOPE project sponsored several national workshops involving hundreds of leading scientists as well as an international summary conference. The publications arising from these workshops were widely read, and though several of the scientists participating in these workshops had indeed read Elton's book it was the SCOPE project that launched modern invasion biology as a scientific field.

The field has since exploded, with its own scientific journals (for instance, *Biological Invasions* and *NeoBiota*), several textbooks, and thousands of scientific papers. For the same reason that scientists became increasingly interested in invasions—the ever-increasing number of environmental and economic problems associated with them—the public also became energized during the 1980s. The great American conservationist Aldo Leopold had called attention to conservation threats posed by introduced species by the 1940s, treating the topic in a major essay in his posthumously published *Sand County Almanac*. However, invasions were not initially a major aspect of the developing conservationist agenda. Occasionally conservation biologists raised the alarm. For example, Raymond Dasmann in 1971 published a manifesto, *No Further Retreat*, to save Florida from environmental destruction; it included a full-blown treatment of the many impacts of introduced species in Florida. However, the conservation community did not become substantially engaged in the issue until the early 1990s, when the Rio Convention on Biological Diversity included a ringing statement, Article 8h, committing nations signing the treaty "to prevent the introduction of, control or eradicate those alien species which threaten ecosystems, habitats or species." An international conference on biological invasions was held in Trondheim, Norway, in 1996, sponsored by the United Nations and the Norwegian government, leading to the establishment in 1997 of the Global Invasive Species Programme, an international effort to aid implementation of Article 8h. Several national efforts soon followed with comprehensive laws, including the New Zealand Biosecurity Act in 1993 and President Clinton's Executive Order 13112 on invasive species in 1999.

1.3 Why are biological invasions important?

In 1857, a French lithographer and amateur naturalist, Etienne Leopold Trouvelot, moved to Medford, Massachusetts. Among other pursuits, he collected animal specimens and donated them to the Museum of Comparative Zoology at nearby Harvard University. The Civil War in the 1860s cut off New England textile mills from their supply of southern cotton, spurring Trouvelot to begin experimenting with native American moths to try to find one that spun fiber for its pupae that could be used as a type of silk. None were productive enough to serve as a basis for a commercial industry, and in 1866 Trouvelot went to Europe in search of moths that would better serve his purpose. He arranged for eggs of the gypsy moth to be sent to him as a more promising candidate, and they arrived in Medford in 1868. He reared them in a rickety cage in his backyard, and in 1869 they escaped, triggering one of the most damaging invasions in American history.

The silk produced by the gypsy moth turned out to be worthless, and in any event the end of the Civil War and renewed availability of cotton led to a great decline in interest in generating a North American silk industry. However, the gypsy moth spread rapidly (partly by the movement of egg masses on vehicles and infested timber), easily escaping from an expensive eradication campaign launched by the state of Massachusetts. By 1934 it was established throughout New England, by 1994 it had spread through the mid-Atlantic states and into parts of the South and Midwest, and it is now established as far south as Tennessee and as far west as Wisconsin as it inexorably expands its range. The caterpillars feed on over 300 tree species, but especially on oak, aspen, willow, larch, and birch, causing massive defoliation, which affects birds, insects, and other forest-dwelling animals. Continued defoliation often causes tree death, and 25–30% of trees typically die in an outbreak region. Particularly susceptible tree species decline in frequency as they are replaced by species less affected by the gypsy moth, again with follow-on effects on animal species that used the declining tree species as food or nesting habitat. Gypsy moth caterpillars also have urticating hairs that cause hives and rashes in some people.

The United States Department of Agriculture employed massive chemical campaigns in a futile attempt to eradicate the gypsy moth, with millions of acres sprayed in the 1950s with DDT, inadvertently causing colossal nontarget damage to other insects and to birds and mammals and even affecting human health. The gypsy moth eradication campaign was a key target of Rachel Carson's 1962 book *Silent Spring*, credited with launching the environmental movement in the United States. Beginning in 1906, another government program attempted biological control (introduction of natural enemies of a pest; Chapter 10.3) and released a parasitic tachinid fly. Although it did not control the gypsy moth, the fly persisted, spread, and now instead suppresses several native giant silk moths of conservation concern (Chapter 10.3). The economic cost of the gypsy moth invasion is staggering. States spend millions of dollars each year to attempt to control gypsy moth spread, and huge amounts of valuable timber are nevertheless lost. In 1981 alone, the estimated cost of this invasion to the U.S. economy was $764 million. And to think that it all began in the backyard of an amateur naturalist in Medford, Massachusetts!

As exemplified by the gypsy moth, biological invasions are important for ecological, economic, public health, and cultural reasons. Although ecological impacts have received the most attention and will be detailed in Chapters 3 and 4, everyone needs to know the many ways biological invasions affect our lives and our environment. They not only endanger native species, but they also cost the world economy billions of dollars annually, devastate agriculture, spread painful and even lethal diseases, and otherwise diminish our quality of life in myriad surprising ways. Along with climate change and massive habitat destruction, biological invasions are one of the great global changes wrought by human activities, and the full consequences are far from understood.

Perhaps the most visible and publicized impact of introduced species occurs when an introduced predator attacks and eats native prey. Photographs of the introduced brown tree snake eating songbirds on Guam not only attracted public attention but also alarmed President Clinton, who vowed

Figure 1.1 16½ foot Burmese python captured in Everglades National Park, Florida, 2012. (Photograph courtesy Catherine Puckett, U.S. Geological Survey.)

to keep the snake from reaching Hawaii. More recently, the Internet quickly spread a picture from Florida of a Burmese python that had swallowed an alligator, which then proceeded to eat its way out of the snake's stomach (Figure 1.1). Rats introduced to islands worldwide have devastated many seabird populations, preying on eggs and nestlings, while the Nile perch, introduced to Lake Victoria, preyed on native fish species, eventually driving at least 200 species to extinction. Introduced animals also eat plants, wreaking havoc on agriculture, forestry, and gardens. The Russian wheat aphid, introduced to the United States in the 1980s, causes massive damage to wheat crops. The sweet potato whitefly, native to Asia, is now on every continent except Antarctica and attacks many crop plants, especially in warm regions, particularly targeting cassava, beans, squashes, and peppers. Introduced European rabbits have devastated vegetation worldwide, especially on islands. Forests in North America have been ravaged by introduced insects such as the gypsy moth, emerald ash borer, and hemlock woolly adelgid.

Introduced species can also compete with native species for resources such as food, light, or nesting sites. The eastern gray squirrel from North America, brought to Great Britain in the early 19th century as a curiosity and released to nature in 1876, has gradually replaced the native red squirrel, partly by foraging more efficiently for nuts. The Asian house gecko has outcompeted native geckos for insect prey on several Pacific islands. European starlings compete with North American sapsuckers for nest cavities. The identities of native species outcompeted by an invader can be surprising. In New Zealand, many native birds that inhabit southern beech forests, such as bellbirds, tuis, and an endemic parrot, the kaka, depend on a rich "honeydew" produced by native scale insects. The invasive northern hemisphere yellow jacket has achieved such high densities that it outcompetes the birds for honeydew, harvesting over 7 pounds per acre, more than all the birds combined. Animal species are not the only ones to compete—plants can outcompete one another for light. For example, South American water hyacinth covers the surface of water bodies in North America, Africa, and Asia and prevents sunlight from reaching plants in deeper waters.

Some introduced species have major impacts not by depleting resources on which natives depend but simply by virtue of aggressive behavior. Several invasive ant species, such as the red imported fire ant and the Argentine ant, greatly reduce native ant populations by attacking them.

Introduced pathogens and parasites can damage native species. Whirling disease, a European parasite of trout, got to North America in frozen trout shipped from Scandinavia to grocery stores in Pennsylvania. Spores from trout remains then reached nearby streams and ultimately a large rainbow trout hatchery. From there, infected fingerlings were shipped to western states, where the disease has now devastated rainbow trout sport fishing. Avian malaria, introduced to Hawaii with Asian songbirds, spread rapidly and contributes to the endangerment of many native bird species. The eastern oyster, devastated by overfishing and sedimentation, suffered a further crushing blow when Pacific oysters introduced to Chesapeake Bay carried two diseases, dermo and MSX. Introduced diseases have similarly ravaged forests. Dutch elm disease, an Asian fungus introduced first to Europe and then, in the 1920s, to the United States, destroyed vast numbers of elm trees on both continents. Agriculture can also suffer huge losses from introduced pathogens; the potato blight, a fungus-like organism from South America, first reached North America, then crossed to Europe in a shipment of potatoes in the 1840s and caused the infamous potato famine in Ireland.

Many of the most ecologically consequential invasions consist of modification of entire habitats, the effects of which then ripple through an ecosystem and affect many of the resident species. Chestnut blight, introduced from Asia, destroyed almost all individuals of what had been the dominant tree species in much of eastern North America during the first half of the 20th century. Its disappearance changed both the physical structure of many forests and the cycling of nutrients such as nitrogen. Over much of its former range, American chestnut was replaced by various oak species. Among other impacts, at least seven moth species that had depended on chestnut went extinct.

Several introduced plant species simultaneously affect many native species by drastically modifying existing fire regimes—the frequency, intensity, and timing of fires. Cheatgrass, from southern Europe and southwestern Asia, was first established in western North America around 1899. Spreading along railroad lines and as a contaminant of commercial seeds, it invaded shrublands by occupying bare soil between shrubs. It burns very readily, greatly increasing the frequency, intensity, and extent of fires. The new fire regime devastates most native western shrubs—such as sagebrush—that are not fire-adapted. When native plants then disappear or become sparse, insects and other animals adapted to the original vegetation decline or disappear.

Invasive plants can also change the hydrology of an entire ecosystem and thus affect many of its inhabitants. In California, Mediterranean salt cedar, by virtue of its deep roots and rapid transpiration, drained the surface water of a large marsh, thereby eliminating much of the associated biota. Common cordgrass is a new species produced in the late 19th century in England by hybridization between the native small cordgrass and smooth cordgrass, introduced from North America. On parts of the southern coast of England and the coast of the state of Washington (where it was subsequently introduced), common cordgrass has transformed intertidal areas and gently sloping mudflats into poorly drained marshes. This transformation in turn has led to great changes in the resident animal community.

Although invasions that modify entire ecosystems are usually by introduced plants or introduced pathogens of native plants, introduced animals can also bring about drastic changes at the ecosystem level. In 1946, North American beavers were introduced to the archipelago of Tierra del Fuego at the tip of South America in an attempt to start an industry based on beaver fur. Although this industry has not thrived, beavers have multiplied enormously. By cutting trees and building dams, they have transformed forests into meadows and also fostered the spread of introduced ground cover plants. Now numbering in the tens of thousands in both Chilean and Argentinian parts of the archipelago, beavers are the target of a binational campaign to prevent them from spreading to the mainland of these two nations. European wild boar, feral pigs, and their hybrids have caused enormous damage to entire ecosystems in many regions, turning up large patches of vegetation by rooting for tubers and mushrooms. This activity causes erosion, especially in mountainous areas, and also affects the plant community and nutrient cycles by mixing the organic and mineral soil layers. Introduced wild boar also have been part of an invasional meltdown, a phenomenon discussed in Chapter 3 in which two or more introduced species combine to produce a far greater impact than any one of them would have caused alone. For instance, in Patagonia, wild boar root for fungal mats of mushrooms called mycorrhizae. These are mutualistic fungi that help plants to gather nutrients, especially in nutrient-poor soil. On Isla Victoria in the Patagonian Andes, many conifer species, introduced nearly a century ago, have barely spread from the plantations in which they were planted, because the spores of their mycorrhizal fungi are underground and no native animals move them. However, introduced boar have begun to seek out these fungal mats, eat them, and disperse the spores in their feces, thereby speeding up a conifer invasion.

Certain biological invasions also incur enormous economic costs. It has been estimated that, in the United States alone, biological invasions cost the economy about $120 billion annually. For the United States, United Kingdom, Australia, India, South Africa, and Brazil together they cost over $330 billion annually. These costs include losses of crops to insect pests, rats, and weeds, losses of livestock to introduced pathogens, and costs of maintaining water systems fouled by introduced mollusks.

Many economic costs of invasions are straightforward to estimate—for example, costs incurred in fighting a particular introduced insect or weed, in terms of matériel and personnel employed. Other costs are not so easily measured—for instance, one can estimate the total amount of a crop lost to an insect or weed, but the market value of the remaining crop may increase by virtue of the loss, which will lessen the cost to the farmer. Economists have developed methods to account for such complications. For example, the cost to the economy of Montana caused by the loss of the rainbow trout sport fishery to introduced whirling disease can be estimated by accounting for the tourism that did not happen because of this loss, including the fishing and other supplies the tourists would have bought.

Other economic costs are still murkier, particularly costs that do not directly affect a market. For instance, what is the economic value of experiencing a stroll through an American chestnut forest, an opportunity now lost because of the introduced chestnut blight? What monetary value can we attach to the continuing survival of Hawaiian bird species threatened by invaders? Economists have tried to estimate such costs, for example by determining the travel costs of people to some natural area as an estimate of their willingness to pay for the existence of a "natural" site or simply by surveys asking people how much they would be willing to pay to avoid the extinction of a Hawaiian bird species. But such approaches remain controversial, as does the underpinning idea of valuing a natural landscape or a species' existence in dollars.

Public health costs of biological invasions can be enormous. West Nile virus, which infects birds as well as humans, causes a type of encephalitis. It first arrived in the United States in 1999 in New York City, probably as a result of a stowaway bird or mosquito transported from the Middle East. Sick people were not immediately diagnosed because this was a new and unknown pathogen in the United States, but a high frequency of dead birds, especially crows, helped alert authorities to the introduction. Many bird and mosquito species vector West Nile virus, and they quickly carried it throughout the United States; it reached California by 2003. The situation was exacerbated by the invasion of an Asian mosquito (*Aedes japonicus*) in 1998, which may have arrived as larvae in used tires or as adults that stowed away on a plane. This species also vectors West Nile virus, and it aided the quick spread of the virus to many states. West Nile virus is costly. A single outbreak in Sacramento County, California, saw 163 people infected at a cost of over $2 million for medical treatment, with another $700,000 spent on mosquito control. Similarly, an outbreak in Louisiana in 2003 infected 329 people and cost over $20 million.

Such invasions by mosquito-borne viruses are increasingly common. In 2006, chikungunya, a viral disease native to tropical Africa and Asia, reached the Indian Ocean island of La Réunion from east Africa, infecting more than one-third of the island's population. The main culprit was another introduced species, the Asian tiger mosquito. Medical costs were estimated to be $60 million, while the cost of lost worker productivity added another $25 million. In 2007 a smaller outbreak of chikungunya struck in Italy, where the Asian tiger mosquito has also invaded. This disease could easily spread if it reaches the United States; the Asian tiger mosquito was introduced to Houston, Texas, in 1985 in used tires and now infests at least 25 states.

Introduced pathogens have ravaged human populations for centuries. The initiation of European exploration in the New World and on distant islands worldwide brought smallpox, influenza, measles, and other diseases to native peoples who lacked immunity. By some estimates, as many as 80% of all native Americans died of such diseases by the end of the 17th century. Only recently have economic costs of these epidemics been estimated.

Biological invasions can also have huge cultural impacts. The chestnut blight that virtually eliminated the dominant tree over much of eastern North America also eliminated a major economy based on chestnuts and chestnut wood, a cuisine based on chestnuts and chestnut flour, and, among rural populations, many traditions associated with harvesting chestnuts. A similar disaster struck Europe in the late 19th century, when British botanists brought American grape vines to Great Britain. A small hitchhiking aphid-like insect, the phylloxera, devastated British wine grape vines, which were not resistant. The insect then spread to mainland Europe, where it ravaged vineyards throughout the continent but especially in France. An estimated 75% of all European vineyards were destroyed, and the industry began to recover only after resistant American rootstocks were used to establish new vines.

Entire villages dependent on wine production emptied, and many traditions associated with grape culture and harvest were lost regionally.

1.4 What controversies surround invasions and their management?

The growing recognition of biological invasions and their impacts, greater attempts to manage them, and increasing publicity about the phenomenon have generated several controversies. These fall primarily into six categories, all explained in more detail in Chapter 11.

Perhaps the oldest controversy related to invasions stems from the argument that complaining about introduced species is a covert form of xenophobia—a hatred or fear of foreigners simply because of their origins. In fact, some terms used to describe introduced species (e.g., aliens, exotic) have also been used by people lamenting the presence of human immigrants. It is easy to see that some of the horticulturists who were antecedents to the native gardening movement had more in mind than just nonnative plants. For instance, the prominent Danish-American landscape architect Jens Jensen, an advocate of using only native plants, wrote this in 1937: "The gardens that I created myself shall...be in harmony with their landscape environment and the racial characteristics of its inhabitants. They shall express the spirit of America and therefore have to be free of foreign character as far as possible...the Latin and the Oriental crept and creeps more and more over our land, coming from the South, which is settled by Latin people, and also from other centers of mixed masses of immigrants. The Germanic character of our race, of our cities and settlements was overgrown by foreign [character]. Latin has spoiled a lot and still spoils things every day."[1] Surely this statement reflects xenophobia, and a number of other statements by some early 20th-century opponents of introduced species can be chalked up to xenophobia.

For Jensen and several other early 20th-century native plant enthusiasts, the antipathy toward nonnative species was also at least partly an aesthetic judgment—they claimed that nonnative plants simply did not *look* right. Nowadays only a small fraction of the concern with nonnative species, especially nonnative plants, is grounded in either xenophobia or aesthetics. Much more is based on the various impacts previously outlined and treated in more detail in Chapters 3 and 4. Furthermore, an aesthetic objection to nonnative species, though surely a matter of taste, is not necessarily grounded in xenophobia. A discussion of the underlying bases for aesthetic judgments is well beyond the scope of this book. Suffice it to say that many philosophers believe that detailed knowledge of nature can generate better-informed aesthetic judgments about natural subjects, such as a landscape or a species. Knowledge of the impacts of introduced species would surely be part of such detailed knowledge. In any event, the overwhelmingly dominant objection to nonnative species is that they can generate many negative impacts, some of which will not occur until long after their introduction (Chapter 4), and such an objection nowadays grows from broad-based concerns that have nothing to do with xenophobia.

Critics of invasion biology sometimes adduce as evidence of xenophobia the fact that occasionally native species have negative impacts similar to those of introduced species, yet there is no widespread hue and cry to eliminate these native species. While it is true that some native species do have harmful impacts, as observed already, this criticism is incorrect on one count and misleading on another. It is misleading because far fewer native species create such impacts, and, when they do, their spread is almost always facilitated by some other human impact, such as grazing by livestock or a changed fire regime. It is incorrect because, in fact, when a native species does cause a significant negative impact, management procedures are often undertaken to control the population and minimize the impact, exactly as for nonnative invaders.

A second controversial matter is the argument by critics of invasion biology that most introduced species do not become invasive or harmful, and attention should be focused rather on the harm species cause and not on their geographic origin. This argument has several faults. For one, we do not really know how many introduced species cause harm, especially ecological harm, because many invasion impacts are subtle, even if they can be highly consequential. For instance, introduced plants that fix nitrogen, thereby fertilizing previously nutrient-poor soil, can entrain a major shift in the nature of a plant community, but one that will take a long time to be manifested. Furthermore, in many cases introduced species remain localized for long periods, then rapidly spread and become invasive, often with substantial impacts. Often such time lags are broken by some environmental change, such as a changed habitat. The reasons for other lags, and their termination, are mysterious, but there is no doubt that they occur; examples are given in Chapter 4. Thus, the number of introduced species currently known to be causing harm is likely to increase. Given the disproportionate number of nonnative species that ultimately do cause problems and our difficulty in predicting which these will be, it is sensible to be wary of introduced species and to try to prevent their spread.

A third controversy over introduced species arises because many of them are useful to humans in some way. For example, the seven leading crop plants in the United States are introduced. No one denies the utility of many nonnative species, although some would argue that native species could serve certain roles at least as well, perhaps at lower environmental cost. For instance, native garden enthusiasts have pointed to native plants that could replace many nonnative horticultural favorites. Many such replacements would be ecologically advantageous—for example, the use of native desert plants in the American Southwest obviates the need for massive amounts of water to subsidize green lawns. However, this controversy about useful nonnatives is misguided, because both policy and management are not devoted to getting rid of wheat, say, or potatoes. Rather, they target nonnative species that have detrimental impacts, and wheat and potatoes do not cause harm (except when we eat too much of them!). Of course, to an extent, "usefulness" is in the eye of the beholder. Even as widely reviled a plant as Japanese knotweed, viewed as an invasive ecological scourge in much of the United States and Canada, has its defenders. Some claim it is useful in treating Lyme disease and rehabilitating degraded habitats.

A fourth controversy is that the native or nonnative status of some species is uncertain, either because their evolutionary history is unclear or because, as with the capercaillie noted previously, a native species has been eliminated locally and then reintroduced from elsewhere. Or, as with the wisent, hybridization with individuals from elsewhere has produced a local population whose designation as native or nonnative is arbitrary. Nevertheless, only a very few populations are problematic in this way—in the vast majority of cases, a species' native or nonnative status is clear. A controversy over invasion biology and management with respect to this issue is therefore a tempest in a teapot. However, in particular cases, such as that of the wisent, well-informed persons can disagree about whether a species is nonnative and whether or not to attempt to eliminate or manage it.

A fifth, persistent controversy is over various methods used to eradicate or control populations of vertebrates, particularly mammals but occasionally birds. This issue concerns animal rights. Some argue that humans have no right to kill an individual of a sentient species, for whatever reason, and that an individual introduced animal is in any event utterly blameless because it was humans who introduced it, or its ancestors; it does not deserve punishment because it did not choose to arrive. Other people, even if they concede the need to eradicate or at least control a particular introduced population, object to various management methods on the grounds that they are inhumane. For instance, in Hawaii rooting and wallowing by introduced wild boar and their hybrids with feral pigs cause heavy erosion. The eroded areas are breeding grounds for introduced mosquitoes that spread introduced avian malaria, to which endangered native birds are highly susceptible. Because of the mountainous, inaccessible terrain, snares are by far the most effective control method—some would argue this is the only effective method. Yet snares are surely inhumane, particularly if snared individuals are not very quickly dispatched—often an impossibility. No resolution of this controversy is imminent. Biologists work hard to produce humane yet effective tools—a particular aim is the development of effective contraceptives that could be delivered in the wild and eliminate the need for killing. However, such research is slow, and major advances are rare. In the meantime, some introduced species are causing long-term and sometimes even irrevocable harm (e.g., extinction of a native species), so managers feel compelled to act.

Finally, some critics of the entire effort to control biological invasions argue that it is futile because the array of forces that lead to invasions—especially increasing international trade and travel—is too great. Therefore, the best policymakers and managers can do is to delay an inevitable deluge of introduced species, an effort that is far too costly to warrant its continuation. Better, in this view, to attempt to mold the "novel ecosystems" that will result from mixtures of native and nonnative species and especially to engineer them so that they produce ecosystem services that humans require, such as flood control and crop pollination. Others have countered that this view is far too pessimistic with regard to our ability to control invasions and far too optimistic regarding our ability to fashion stable ecosystems that provide the multitude of services we require. Also, advocacy of novel ecosystems and abandonment of the attempt to control invasions consigns a certain number of native species to extinction, some in the near term, others perhaps in a few centuries. Some may view such a loss as inevitable anyway, and perhaps not too tragic. A discussion of these issues is well beyond the scope of this book. Suffice it to say that much of the field of conservation biology is dedicated to the proposition that saving species is not only feasible but also worthwhile.

2

MAGNITUDE, GEOGRAPHY, AND TIME COURSE OF INVASIONS

2.1 How many introduced species become invasive?

To become invasive, a nonnative species must first arrive and establish a population. It is difficult to determine how many species actually arrive, because many of these do not survive or, if they survive, often do not establish populations that persist more than a short while. Now, if the new arrival is a showy bird, there may be a decent chance that someone will observe it before it disappears, but for the vast majority of arrivals with transient existences no one will know they were ever there. Consider the likelihood of observing one or a few individuals of an inconspicuous plant or insect, for instance. For a few types of introductions, such as those of insects used in biological control of pests (Chapter 10.3) or attractive birds introduced by acclimatization societies (Chapter 6.3), good records exist of which species were introduced to compare with which ones established populations and which became invasive. However, the very fact that someone wanted to establish populations of these species probably means that they took pains to see that the introduction succeeded, and that would mean trying to introduce large numbers of individuals and releasing them in an appropriate habitat at the right time of year.

Most arrivals, however, got to their new homes on their own, probably in very small numbers and with no guarantee of landing in a suitable habitat. Unfortunately, this means that records of biological control and acclimatization society introductions, though interesting and useful for identifying certain types of patterns, are probably not too helpful for determining what fraction of introduced species actually became invasive.

Chapter 5.1 will describe in more detail how occasionally a widespread invasion can begin with just a few individuals—for instance, the muskrat in Europe and the house sparrow in North America. However, these "success" stories are surely the exceptions, and the great majority of introductions involving a limited number of individuals are probably failures, not even establishing reproducing populations. They disappear without any record of their presence. Trying to quantify what fraction of introductions establish populations and what fraction of those go on to become invasive has proven challenging. Ecologist Mark Williamson and colleagues in the 1980s and 1990s proposed a rule of thumb, the "tens rule," suggesting that about 10% of introduced species establish populations and 10% of those go on to become *pests*. Pests are just one subset of *invaders*, as explained in Chapter 1.1, but this rule has been a starting point for attempts to determine how many introductions become invasive. Williamson intended the rule to have wide confidence limits around 10%, from perhaps 5% to 20% in each of the two steps.

Some data sets seem to follow the tens rule quite well, at least where data are available to estimate the percentages. For example, of established introduced plant species in the states of California, Florida, and Tennessee, 5.8%, 9.7%, and 13.4%, respectively, became sufficiently invasive to have reached natural areas. Other data sets suggest larger percentages at one or both steps. For instance, ecologist Jonathan Jeschke, looking at mammal and bird introductions, found that 79% and 50%, respectively, established populations; of these, 63% of mammals and 34% of birds became invasive. However, it is quite possible that, even in these relatively visible groups, some failures were not recorded and Jeschke's establishment figures are too high. What these data sets do suggest, however, is that invaders are far from a tiny fraction of all introductions.

2.2 How many biological invasions are there?

No definitive tally exists for any nation or region of the number of introduced species and how many of these went on to become invasive. However, for some well-studied groups of species, estimates are possible. For instance, Great Britain has about 800 established introduced insect species, compared with roughly 20,000 native insect species. Analogous numbers for plants in Britain are about 2,000 established introduced species and about 2,300 native species. For freshwater fishes, the numbers are about 12 established introduced species compared with 38 native species. Established introduced species—that is, naturalized species, reproducing without human assistance—are not all invasive because they have not all spread widely. However, it is likely that between 10% and 20% of the established introduced species are invasive in Britain. Thus, as fractions of the native British biota, introduced insects constitute about 4% of the species, introduced plants about 85%, and introduced fishes roughly 32%. Out of these percentages, we can estimate that between 10% and 20% are invasive.

For the continental United States, an estimated 6,000–7,000 introduced species of plants and animals are naturalized, compared with about 200,000 native species. Perhaps 10% to 20% of these naturalized species are invasive, as in Britain. For certain regions, however, the proportion of introduced species is much higher. In Florida, a particularly heavily invaded state, 27% of plant species, 8% of insects, 29% of land snails, 24% of freshwater fish, 5% of birds, and 24% of land mammals are introduced. The Hawaiian Islands are even more heavily invaded. About half the plants, 25% of the insects, most freshwater fish, and 40% of the birds are nonnative. Islands in general, especially remote ones, have few native species but often many introduced ones. For instance, the Kerguelen Islands, a group of small, isolated subantarctic islands administered by France in the southern Indian Ocean, have a mere 29 native vascular plant species and only about 50 native insect and spider species. However, the islands are also home to at least 68 nonnative plant species and 29 naturalized insect and spider species, 6 of which are clearly invasive.

Recent development of rapid and relatively inexpensive nucleic acid sequencing methods has led to a plethora of genetic research suggesting that many more invasions have occurred than had been suspected and that many invaders were so similar morphologically to native species that no one realized a new species was present. Such "cryptic" invasions may well force us to revise upward estimates of numbers of invasions, especially for groups such as some insects and plants as well as many marine organisms whose taxonomic relationships have not yet been well worked out. A good example is the mussel genus *Mytilus* that inhabits bays in many parts of the world. All mussels in the genus were long thought to be one species, the blue mussel. However, the use of new molecular genetic techniques has now shown that what had been thought to be a population increase of the blue mussel in California was actually an invasion of the Mediterranean mussel, which had not been recognized as a distinct species. The same mussel has now been shown to have invaded Japan. Colonial jellyfish in the genus *Cordylophora* are another example of a recently detected cryptic invader. Several species, all native to the region of the Caspian Sea, are found in various other parts of the world, probably carried in historic times by ballast water. Often these have been viewed as one species because of their morphological similarity. However, molecular differences suggest that they are distinct species, and several sites have been invaded by more than one of these newly recognized species.

2.3 Where do most biological invaders come from, and where do they go?

Since the Columbian Exchange, many Eurasian species have invaded other parts of the world, often with devastating impacts. Although "Exchange" implies that species moved from the New World to the Old World and vice versa, it has long been noted that invaders traveling from Eurasia to North America as well as to many islands worldwide (e.g., the Canary Islands, the Mascarenes, New Zealand) have established and become invasive and often damaging far more frequently than species traveling in the opposite direction. For example, many pathogens—smallpox, measles, typhus, diphtheria, influenza, among others—ravaged other human populations in other areas following their initial contact with Europeans during three centuries of intensive exploration. The only New World disease to devastate Old World populations during this period was syphilis. Similarly, most introduced pathogens that have devastated native plants, such as chestnut blight and Dutch elm disease, have originated in Eurasia and wrought havoc when they arrived elsewhere. However, within the last few decades this pattern has begun to break down.

The reason behind the preponderance of invasions by Old World species into the New World is a subject of much discussion, but there is no doubting the general pattern. As an extreme example, 47 species of European ground beetles in the family Carabidae have established populations in Canada, while not a single Canadian carabid beetle has managed to establish a population in Western Europe. Some fraction of this pattern is probably due to a history of copious opportunities for movement from the Europe to elsewhere. Much of the world's recent history consists of the spread of European civilizations throughout the globe, and it is not surprising that Europeans also spread their animals and plants (and, inadvertently, their diseases) along with their culture. Environmental historian Alfred W. Crosby developed this hypothesis, which he termed *ecological imperialism*. Tracing the statistics and sorrowful stories of many indigenous people in North America, Australia, the Canary Islands, and elsewhere, he charted how European diseases of humans, plants and animals, as well as weeds and domestic animals (some of which became feral in their new homes) spread throughout the world. Crosby viewed the various Old World species as "biotic allies" of European settlers—together, they constituted a juggernaut that crushed native peoples and destroyed their ecosystems, the newly introduced pathogens, plants, and animals being deadlier than guns.

But why should these various European species have been able to dominate and replace the native species in so many places? Why did the native species not instead crush the invaders? A part of the explanation may be that European plants had evolved alongside cattle, sheep, goats, and pigs and thus had evolved means of coexisting with them. By contrast, native plants in North America and elsewhere lacked these adaptations, so introducing livestock favored these European weeds. Evidence from field experiments in North America supports this notion.

The impact of coevolution with humans on dominance by introduced Eurasian species was fleshed out by biologist Francesco Di Castri, who invoked two key forces. First is Europe's long substantial human history. Human impacts are detectable in Europe for at least the past 40,000 years, and strong human impacts on the environment have been manifested for at least the past millennium. The second feature of Di Castri's theory involves both the geography of Eurasia and some climatic history. Eurasia is the largest continuous landmass, and also the extent and effects of glaciations over the past 2.5 million years were greater in Eurasia than elsewhere. But the size and the roughness of the Eurasian landscape favored the existence of several large refugia that allowed many species to persist when glaciers covered much of the land. The many species restricted to refugia were forced to interact both with one another and with humans and also had to adapt to great climatic vicissitudes. In various ways, after being continually subjected to these pressures, Eurasian species may have evolved to be "stronger" in some sense (thus able to overcome species of other areas when they were introduced) and to coexist with humans.

Although Di Castri's theory is logical, it is far from proven. One obvious alternative explanation for at least part of the dominance pattern, one that perhaps could have acted in concert with the forces Di Castri posits, is propagule pressure. A propagule is a group of individuals that can potentially reproduce and thus propagate the species, especially in a newly colonized site. It could be that over the relevant time period, beginning with the Columbian Exchange, many more species moved from Eurasia to the rest of the world than in the opposite direction. Some data from the history of European colonization of the New World would seem to support this view. For instance, a large fraction of the first Old World insects to be recorded in North America were beetles that typically live in soil and whose native range includes southwestern England. As it happens, this area is the last point of land where early ships sailing to America stopped before crossing the Atlantic. It was here that they picked up soil to carry as ballast in the ocean, unloading the ballast once they reached North America. So it seems obvious that the reason for so many early introductions of European soil-dwelling beetles to North America is the propagule pressure generated by the amount of soil, and the beetles living in it, transported from Europe.

However, it is equally clear that increased propagule pressure produced by hitchhiking cannot explain the entire imbalance between Old World invaders in the New World and New World invaders in the Old World. After all, many of the ships that carried soil ballast to America and then dumped it there then took on cargo, such as timber, that should have provided ample opportunity for New World hitchhikers, yet relatively few such species invaded Europe. The disparity between

Old World and New World invaders remains a conundrum. Certainly there are exceptions—New World species that have heavily invaded the Old World with substantial impact on the natives—for example, the aforementioned replacement in Great Britain of the European red squirrel by the North American gray squirrel. Yet another example is the signal crayfish, native to the Pacific Northwest of North America, imported in the 1960s to Scandinavia to support sport fishing and commerce. It then quickly spread as far as Greece, Spain, and Great Britain, attacking the more passive native crayfish and spreading a crayfish plague, to which it is resistant and the European crayfish are susceptible. In 1973, a second crayfish species, the red swamp crayfish, native to North America from the Gulf coast to the southern Midwest, was also introduced to Europe. In southern Europe, it is now replacing native crayfish by outcompeting them and spreading crayfish plague (Figure 2.1). In wetlands, the North American muskrat, introduced to Europe in the early 20th century, has spread widely, and its herbivory has had many impacts on native plants and

Figure 2.1 North American signal crayfish standing on introduced zebra mussels in Sweden. (Photograph courtesy Anders Asp.)

animals. Among plants, black locust, black cherry, fleabane, and ragweed are common, damaging invaders in Europe, but no North American plant dominates large areas of Eurasia to the extent that Eurasian cheatgrass and spotted knapweed do in North America. Still, there is no guarantee that New World species will not continue to invade the Old World, and the basis of the disparity to date in invasions between the two regions remains largely unexplained.

Globally, several other invasion patterns have been noted. Boreal forests of North America and Eurasia, dominated by coniferous trees and lying between the tundra in the north and deciduous forest in the south, have suffered few consequential invasions. No single species of introduced tree has yet transformed large expanses of boreal forest to the extent that tree invasions have modified temperate and subtropical forests (Chapter 3). Several northern European trees, such as Scots pine, have been planted in boreal North America, but all have stayed close to where they were planted or at most have colonized only areas heavily disturbed by humans. Neither have boreal forests yet been devastated by introduced plant pathogens such as the chestnut blight or Dutch elm disease or insects such as the emerald ash borer. In northern Sweden, although over 1,400,000 acres were planted with North American lodgepole pine, particularly in the 1970s and 1980s, the species has not yet spread substantially from its many plantations. Neither have other Eurasian tree species planted in Canadian forest or North American trees planted in northern Europe become prominent invaders. Of course, there is always the fear that one or more of these is simply in a lag phase (Chapter 4.3) and will ultimately become invasive. The forest industry, by contrast, worries that a plant pathogen such as chestnut blight will ultimately arrive and devastate widespread nonnative plantings such as lodgepole pine in Europe.

Few consequential invasions have been reported to or from lowland tropical rain forests of Africa, Asia, South America, and Australia. Similar habitat on islands has been devastated by introduced species, as have many parts of many islands worldwide (Chapter 2.5). However, the large continental rain forests are largely unaffected. Part of the reason may be that few rain forest trees have been introduced to nonnative regions compared with temperate and boreal trees. However, disturbances in tropical forests, such as land clearing and road building, often lead to invasion by both plants and animals in the disturbed areas, some of which may not be able to return to their original state without substantial restoration activities. For instance, the earthworm *Pontoscolex corethrurus*, native to the Guyana plateau in South America, has become a problem in other parts of South America and in Queensland rainforests in Australia. It proliferates rapidly in the wake of logging, land clearing by fire, or road construction; displaces native worms; and changes the soil to the detriment of at least several native plants. At some sites, this worm compacts the soil much as bulldozers would and changes soil chemistry.

Australia's history is laden with invasions—some legendary, such as those of the rabbit from Europe, the cane toad from the Americas via Hawaii, and the prickly pear cactus from the Americas. Each of these species was deliberately introduced for one purpose or another (Chapter 6). Australian biologist Tim Low argues that Australia is currently in the process of a global payback. He points out that, in many regions of the world, Australian plants are now dominating large parts of the landscape and causing wholesale ecosystem change. Among better-known examples are paperbark and Australian pine in Florida, eucalyptus in areas of California and South America, and silky hakea, black wattle, and coojong in South Africa. As Low observes, much of Australia is dry, so many of its plants are adapted to dry conditions—conditions that will increase in many regions as global climate changes. However, this fails to explain some of the Australian invaders, such as paperbark, that are not invading dry areas. Possibly the surge of these species in various parts of the world simply reflects increasing propagule pressure brought about by increasing world trade.

In sum, the sites and sources of invasions can often be explained by the history of human activities that move species. However, an apparent propensity for Eurasian species to invade elsewhere cannot be wholly attributed to history. Other patterns, such as the relative dearth of invasions into boreal forests and moist tropical forests, are not easily explained and may be only temporary. A final dominant pattern—frequent devastating invasions on islands—will be detailed in Chapter 2.5.

2.4 When have invasions occurred and by what means? How have rates of invasions changed?

As will be discussed in more detail in Chapter 11.3, biologists focus primarily on invasions that have occurred since the beginning of global exploration and settlement by Europeans 500 years ago. This is largely because most invasions have occurred since then, though some happened earlier. As an example of a very early invasion, the Pacific rat, pigs, and other plants and animals were spread through the Pacific islands by the Lapita people as early as 3,000 years ago. Excellent evidence also exists for the early introduction of mammals to the Mediterranean island of Corsica. Paleobiologist Jean-Denis Vigne has demonstrated through fossil remains that sheep, goats, and pigs were brought to Corsica by humans about 7,000 years ago, and cattle, dogs, and foxes slightly later. All but foxes were surely deliberately introduced. The European hedgehog, field mouse, and dormouse first arrived between 6,000 and 5,000 years ago, the first two as stowaways and the dormouse possibly as a deliberate import. The house mouse, garden dormouse, lesser white-toothed shrew, and Etruscan shrew (the world's smallest mammal) all hitchhiked to Corsica by 2,500 years ago, and possibly earlier. Horses and donkeys were also introduced around 2,500 years ago, and the Corsican wildcat is probably a descendant of domestic cats introduced at that time. Red deer were brought at least by 1,600 years ago, and ship rats first arrived at least as early as 600 AD. In fact, the current mammal fauna of Corsica (more than 25 species) is wholly introduced. Extinction of the original mammals (at least seven species, including a fox and a large rabbit) was caused by hunting, predation by the introduced dog, fox, and later the weasel, and possibly competition with some of the invaders.

With European settlement of many regions after Columbus, some invasions took place remarkably quickly, as evidenced by early reports such as Pehr Kalm's observation of Old World plant species in North America by 1748 (Chapter 1.2). Darwin, traveling through the Argentinian pampas in the first half of the 19th century, was astounded by the degree to which two introduced thistles, aided by introduced cattle, already dominated vast acreage: "There were immense beds of the thistle, as well as of the cardoon; the whole country, indeed, may be called one great bed. The two sorts grow separate, each plant in company with its own kind. The cardoon is as high as a horse's back, but the Pampas thistle is often higher than the crown of the rider's head. To leave the road for a yard is out of the question; and the road itself is partly, and in some cases entirely closed." Speaking of the cardoon, Darwin added, "In the latter country alone [part of Argentina], very many (probably several hundred) square miles are covered by one mass of these prickly plants, and are impenetrable by man or beast. Over the undulating plains, where these great beds occur, nothing else can now live....I doubt whether any case is on record of an invasion on so grand a scale of one plant over the aborigines."[1] The pampas thistle to which Darwin referred, now known as the milk thistle, is native to southeast Europe and Asia, while the cardoon is from the Mediterranean region. This is a perfect example of Crosby's notion, cited previously, that species that had adapted previously to humans and their livestock can act in concert with one another to crush native biological communities.

Many different examples could be adduced of particular regions of the earth being progressively invaded by various species over the last few centuries, the details of tempo and scale determined by the history of human activity and the particular type of species. The North American Great Lakes are a particularly well-studied case. They currently contain thriving populations of at least 185 nonnative species of fishes, invertebrates, plants, pathogens, and algae, none of which were present before 1800 and all of which either were introduced deliberately by humans or arrived as by-products of human activities. A major factor was the completion in 1825 of a network of canals linking the Great Lakes and the Atlantic Ocean. These canals allowed species that had formerly been separated by land barriers to reach new drainages. For example, the Piedmont elimia, a small snail from the Hudson River drainage, soon reached the Great Lakes through the Erie Canal. The Saint Lawrence River canal system, opened in the mid-19th century, allowed more Eurasian species entrance to the Great Lakes.

The solid ballast used by 19th-century ships was one route by which organisms could arrive; another was by fouling of the ships' hulls. The sea lamprey, which caused the extinction of at least three native fish species (Chapter 3.3), reached Lake Ontario by the 19th century, probably on a fouled hull, and then spread to Lake Erie by 1921 with the completion of the Welland Canal (which bypassed Niagara Falls). It then traveled through the other lakes via the canal system. The alewife, a small Atlantic fish that became enormously numerous and devastated zooplankton populations (Chapter 3.2), reached Lake Ontario by the mid-19th century and proceeded to spread into Lake Erie by the Welland Canal. By 1954 it could be found in all the Great Lakes. Watercress was introduced as food and by 1847 had spread to Niagara Falls. By 1870, 17 introduced species, including plants, fish, and mollusks, had been introduced to the Great Lakes; as these were the only groups surveyed at this time, other species likely arrived by then as well, and some are probably still not recognized as introduced. The aquarium trade was another route for introductions to the

Great Lakes beginning in the late 19th century. For instance, the banded mystery snail from the Mississippi drainage arrived in dumped aquarium contents by 1910.

In the 1870s, government agencies began introducing many fish species to the Great Lakes, some of which, such as common carp, have become highly damaging invaders. Other fishes introduced to establish fisheries by the 1920s include brown trout from Europe and chinook salmon and rainbow trout from other parts of North America. During this same period, mosquito fish were introduced for mosquito control. Mollusk introductions are believed to have begun by the mid-1800s but were unrecorded.

In the 1880s, technological advances allowed ships to start using water instead of soil as ballast, and this switch eventually precipitated a wave of new invasions. By the early 20th century, water was the normal ballast, and improvements in the Saint Lawrence River canal system as well as the opening of the Saint Lawrence Seaway in 1959 permitted enormous amounts of ballast water from other regions to be emptied into the Great Lakes. In the early to mid-1980s, many famous invaders arrived in discarded ballast water: zebra and quagga mussels from Russia and the Ukraine; round and tubenose gobies; the spiny water flea; and the ruffe from Eurasia.

Stocking of fish has continued to the present day. Coho salmon were introduced in 1933, kokanee in 1950, and pink salmon in 1956. Rudd, sold as bait fish, were released into the Great Lakes by 1989. A developing threat, discussed at the end of this chapter, is the potential arrival of Asian carp species, especially bighead and silver carp, into the Great Lakes from the Mississippi drainage to which they were introduced. Recently, a disastrous introduction of a fish pathogen known as viral hemorrhagic septicemia (VHS) has killed lake trout, steelhead trout, chinook salmon, yellow perch, gobies, emerald shiners, muskies, whitefish, and walleyes. Originally in the Atlantic, it was first recorded in 2003 in Lake St. Clair (which connects to Lake Erie and Lake Huron) and then in

Lake Ontario in 2005; it is now widely dispersed in the Great Lakes and has caused massive die-offs of several fish species beginning in 2006.

The timing of insect invasions in the United States has also been studied in detail. Several insect species probably arrived with Columbus and others with the *Mayflower* and maybe even with early Norse explorers. Early invaders in the Northeast included species closely associated with humans, such as the bedbug, head (and body) louse, and Oriental cockroach. Up until 1800 there were only about 30 introduced insect species in the continental United States, probably because voyages to America from Europe took so long that only those species that could survive and perhaps reproduce on humans or domestic animals, or in stored products or soil ballast, would survive the journey. Almost all insect species that invaded before 1800 were from Europe, the origin of most ships traveling to North America at that time. As mentioned earlier, a large fraction of the early invaders were beetles from southwest England that lived in soil, because that is where soil ballast was often taken on board for the transoceanic voyage (see also Chapter 6.6).

After the American Civil War, steamships greatly sped up the ocean crossing, and nursery stock imported into the United States, especially from Europe, drastically increased the number of introduced insect species. By 1880, there were roughly 120 introduced insects in the United States, and homopterans (plant pests such as scale insects, aphids, and whiteflies) that had arrived on the plants approximately equaled the number of beetles. In addition, plant-dwelling lepidopterans (moths and butterflies) became more numerous, as did hymenopterans (wasps, bees, and ants; some of the wasps were parasites of other hitchhiking insects). The United States Department of Agriculture even housed an Office of Plant Introduction at this time, whose raison d'être was to find "useful" plants overseas and bring them back to the United States. They introduced over 200,000 species and varieties, initially with very little concern for what might be hitchhiking on the plants. By

1920, the number of introduced insect species had ballooned to about 580, with homopterans and beetles still dominating and lepidopterans and hymenopterans trailing closely behind. Around 1920, the rate of increase tailed off somewhat, especially for homopterans, probably owing to stricter application of the previously unenforced Plant Quarantine Act of 1912. The rate of increase of beetles and especially hymenopterans still increased, however. By 1940, there were about 800 introduced insect species, and hymenopterans were as numerous as beetles and homopterans. The hymenopteran increase, and to a lesser extent that of beetles and also flies, reflected the increased deliberate introduction of parasitic wasps and flies as well as predatory beetles for biological control of agricultural pests (Chapter 10.3). By 1980 the United States had about 1,350 introduced insect species, by 1990 about 1,850, and today the estimate is well over 2,000. This increase is doubtless due to the greatly increased rates of cargo transport over this period (Chapter 12.2). What is more, whereas most of the introduced insects in the United States through the 19th and first half of the 20th centuries were European, many have arrived more recently from other continents, especially Asia, reflecting diversified global trade patterns.

The timing of introduction of songbirds and doves to the Hawaiian Islands is also well-known, as are the details of those avian introductions that either failed or died out after establishment (unlike for most introductions of insects to the United States or aquatic species to the Great Lakes). The only introduced bird present in the archipelago in the early 19th century was the Old World rock dove. Between 1860 and 1900, another nine species arrived, including one (the common myna) that was deliberately introduced to control insect pests as well as a few cage escapes or releases (for instance, the nutmeg mannikin and the red munia). Only four introductions of birds occurred between 1901 and 1920, but a burst of activity by acclimatization societies (Chapters 2.5 and 6.3) brought another 30 species from all over the globe by 1940, mostly colorful species that residents thought would enliven their environment. Restrictions on deliberate introductions beginning in 1945 ended the practice of deliberate introduction, so only three species arrived between 1941 and 1960. However, since then the increase in the human population and the proliferating pet trade have led to at least 21 cage escapes or releases (though few after the 1970s). Not all of these birds survived; some probably did not even establish ongoing populations, but others persisted for decades before disappearing. Of at least 69 songbird and dove introductions, 38 species survive today, some with staggering impacts on native birds by virtue of vectored diseases (Chapter 3.5). Many bird species aside from songbirds and doves were introduced to Hawaii (for instance, North American mallards and Old World pheasants for hunting); all told, at least 170 species have been introduced, and at least 54 have established populations.

These examples of aquatic invaders, insects, and birds exemplify two important generalizations about the timing of biological invasions. First, the details differ depending on the reasons for the introductions and the mode of transport. Second, all of them show a strongly upward trend, with spikes and temporary lulls determined by the specifics of the motives and means of introduction.

2.5 Why are island ecosystems especially vulnerable?

Islands are famous for both having many introduced species and suffering huge transformations at the hands of these species. For example, Hawaii has about 2,500 introduced insect species and only 5,000 native species, while the contiguous 48 states have about 3,000 introduced insect species and 96,000 native insect species. Islands are also well-known for having few native species: they have fewer species than equal-sized areas of equivalent mainland. Charles Elton related the high number of introduced species to the low number of native species in his biotic resistance hypothesis—the presence of fewer native species on islands means that a newly introduced species is confronted with fewer natural enemies (competitors, predators, herbivores, parasites, and pathogens) than on the mainland. One version of this hypothesis states that there are more "empty" ecological niches on islands, so a newly introduced species is more likely to find some way of making a living there that differs from that of any native.

Many data are at least superficially consistent with the biotic resistance hypothesis. For instance, about 3 times as many introduced bird species and 1.6 times as many introduced mammal species have been able to establish persistent populations on islands than on continents. However, without knowing how many attempted introductions were made and failed on islands and continents, one cannot be certain that the island communities were really more easily invaded than mainland ones. For that we would need to know that a higher proportion of introduced species survived on islands. Yet data on failed invasions are scarce, except for bird species (often introduced by acclimatization societies) and insects introduced for biological control (Chapter 10.3). For the latter, the data are equivocal—there is little difference between the fraction of introduced biological control species that survived on islands and mainland. For birds, the data have never been examined in such a way as to answer this question, although it is evident that many islands (e.g., Hawaii, New Zealand, Tahiti, the Mascarenes) have received enormous numbers of introduced birds by persons who colonized the islands in recent times from continents. These settlers then formed acclimatization societies, seeking to fill the "void" created by the absence on the islands of bird species from home. For instance, to return to the Hawaiian islands, at least 69 species of passerine birds (perching birds), pigeons, and doves have been introduced at various times, of which 38 species persist somewhere in the archipelago. A birdwatcher can stroll through the campus of the University of Hawaii, on the outskirts of Honolulu, and easily see a dozen species from five continents (but only one native species). However, 35 of the introduced species did not survive on at least one of the six major Hawaiian islands; 31 of these survived on none of them. These birds were brought from Asia, Europe, Africa, North and South America, and Australia, and in addition a number of other birds—game birds and parrots—were introduced. Thus, at least part of the reason islands have so many species of introduced birds is undoubtedly propagule pressure—people introduced many of them, often repeatedly.

Introduced species on islands have often had staggering impacts. On the Hawaiian Islands, Society Islands, and other small islands in the Pacific, the rosy wolf snail, introduced from Florida and Central America in a futile attempt to lower populations of the introduced giant African snail, has caused the extinction of at least 50 species of terrestrial snails. The small Indian mongoose, introduced in attempts to control rats in agricultural fields and occasionally to control venomous snakes, has caused extinction and endangerment of mammals, birds, reptiles, and amphibians in the West Indies, Hawaiian Islands, Fiji, Okinawa, Amami, Mauritius, and several Adriatic islands. Feral cats have plagued birds on many small islands. For instance, on subantarctic Marion Island in the southern Indian Ocean, cats were introduced in 1949 and are now estimated to kill over 400,000 birds per year, mostly ground-nesting petrels. Habitat destruction by grazing rabbits eliminated two endemic birds from the tiny Hawaiian island of Laysan. Some plant invasions have also devastated islands. For instance, the Central American bush currant tree, introduced to Tahiti, now heavily infests 60% of the island, has caused extinctions of several plant species, and threatens many more.

Introduced species have also carried nonnative diseases that have devastated susceptible island species. For instance, in the Hawaiian Islands, introduced mosquitoes have transmitted introduced avian pox and avian malaria to highly susceptible native birds, contributing to the extinction of several species and the threatened or endangered status of others. Many native people on islands have also been ravaged by introduced diseases beginning with the Columbian Exchange.

Many invasions have, of course, also affected native species on the mainland. However, on average, their impact is less on the mainland than on islands. For instance, introduced species are at least one cause of endangerment for 16 of 19 endangered Hawaiian bird species but for only 2 of 13 endangered mainland species. The worldwide statistics on endangered birds paint a similar picture: of threatened island species, introduced species are the sole threat or one of the threats for 38%, while they are a factor for only 5% of threatened mainland bird species. Similar data are available for threatened plants of the California mainland and the California Channel Islands. For 19 threatened island plant species, introduced species are a factor for 17 of them. Of mainland species, only 21 of 134 are threatened by introduced species.

Several writers have claimed that the reason island ecosystems are so devastated by introduced species from the mainland, and why such a large fraction of island species have gone on to become extinct or are endangered relative to mainland species, is that island species are weaker in some sense than mainland species. Carol Kaesuk Yoon, speaking of the Hawaiian Islands, said, "The isolation of these gentle islands from mainland predators and diseases created a flora and fauna ill-equipped to handle the rigors of competition with the outside world. Hawaii is home to thistles without prickles, blackberries without thorns and many flightless insects and birds."[2] Similarly, Frank Preston termed island species "evolutionary backwaters and dead ends."[3] These claims play into the biotic resistance hypothesis for island invasions—not only do islands have fewer species, but also those they do have are weaklings.

However, a closer look suggests that the entire phenomenon of island species endangerment and extinction as well as island community invasibility does not arise simply because island species are inherently weaker or less adapted to their environments than mainland species are. Occasionally an island species seems to outcompete continental ones, as with the New Zealand mud snail that is currently wreaking havoc with the native snail community in the Greater Yellowstone region of North America, and the New Zealand flatworm, a predator greatly reducing many native earthworm populations in Great Britain and Ireland. But even if island species are not all maladapted weaklings, the relatively small sizes of islands and their populations act against island species when mainland species invade their homes.

The small size of islands exacerbates the impact of habitat destruction and transformation. The Hawaiian Islands, the Mascarenes, and many other islands where a plethora introduced species have prevailed over native species have had almost their entire lowland forests cleared for agriculture, housing, and other human enterprises. The species that now inhabit those islands are adapted to such habitats. For example, at least 300 of the world's 1,186 threatened bird species are island forest birds, and, while there is no comprehensive study of what fraction of forest habitat has been destroyed on mainland compared to islands, surely that fraction is much greater on islands. For example, in Hawaii, about a third of forest remains, almost all in high elevations. A total of 14 of 38 forest birds are already extinct since European colonization, and 15 are on the federal endangered species list; only 9 are not endangered. Compare these statistics with the eastern United States, in which only about 1% of virgin forest remains but only 3 of 70 forest species have gone extinct (and two of those did so because of hunting rather than deforestation). However, in the eastern United States, it took over 200 years for the virgin forest to be cleared, and during this time significant portions of the forest remained or were regenerating. At no point was there less than ca. 45% of the original total forest area forested, so there was always a substantial amount left in the range of almost all forest species. By contrast, in

Hawaii, there has been little regrowth, and there was no place left for most of the declining species to go because they had very small ranges to begin with. Hawaiian birds were not driven to extinction because they were "weaker" than those of the eastern United States, nor was it competition with newly arrived invaders that eliminated them. Rather, it was just that the Hawaiian birds had far smaller ranges and no refuges.

Consider also the differing effects of cyclones and hurricanes on islands and continents. The Hawaiian Islands, the Mascarenes, and the West Indies are all in cyclonic pathways, and recent storms have extinguished several bird species and endangered several others. In one instance, a hurricane eliminated a bullfinch on St. Kitts, and another killed almost all Laysan teals. Hurricane Hyacinthe killed half the individuals of several endemic, threatened bird species on La Réunion in 1980. Most striking is Hurricane Iniki, which traversed Kauai in 1992 and extinguished five bird species or subspecies.

These storms had such disastrous impacts on birds because islands are all small and the populations had already been reduced in size and range by habitat destruction. Four of the five birds that went extinct on Kauai had fewer than ten individuals when the hurricane hit. The Laysan teal had already been eliminated from one island and reduced on Laysan, which is only about 1,000 acres. On La Réunion, only about one-fourth of the forest still remains, and only a few hundred acres at most in the lowland forest where these birds live. Surely all these species, having evolved in hurricane belts, would have been able to survive these storms if their ranges had not been so reduced by humans. Probably mainland species, if their ranges had been similarly reduced, would have been devastated just as much; whether or not the island birds were weaker had nothing to do with it. The impact of Hurricane Hugo in 1989 on the Puerto Rican parrot and the red-cockaded woodpecker is instructive. This hurricane eliminated about half of the only population of the parrot, located in Luquillo Forest in Puerto Rico. The parrot was already endangered because it had progressively been restricted to this one site; in fact, when the hurricane struck, a plan was afoot specifically to lessen the threat of such a catastrophe by establishing a second population. So the impact of the hurricane threatened the very existence of this species. By contrast, this same hurricane devastated the Francis Marion National Forest in South Carolina, home to one of the largest populations of the federally endangered red-cockaded woodpecker. The storm did not impact the woodpecker as it did the parrot. The Marion Forest is eight times larger than the Luquillo Forest, and the Marion Forest red-cockaded woodpecker population is just one of six populations; thus, the species could not have been eliminated by a single large storm.

Introduced diseases often devastate island species, and some of these are vectored by introduced continental species resistant to these diseases. A prime example can be found in the decline of several native Hawaiian bird species. Originally, habitat conversion reduced all their populations and relegated them increasingly to upland forests. However, their decline was greatly exacerbated by avian malaria and avian pox introduced with Asian songbirds and vectored by previously introduced mosquitoes (Chapter 3.5). Although continental ecosystems have also been ravaged by introduced pathogens (Chapter 3.5), the smaller size of island populations renders them exceptionally susceptible. Because they are small, island populations are less likely to possess genotypes that are resistant to a new pathogen.

A second factor contributing to the devastating impact of some island invasions is that many islands, especially small ones in the oceans isolated far from major landmasses, lack certain sorts of species. Entire ways of making a living are largely absent on isolated islands. In particular, there are no predatory or grazing mammals. Thus, island species have not evolved defenses against such species, and when predators or grazers are introduced native species suffer enormously. Introduced rats, cats, and the small Indian mongoose, for example, have eliminated many island birds as well as some mammals and amphibians. The impact of the mongoose alone, and just on mammals, is striking: it has contributed to the extinction of four shrew species on Haiti and the dwarf hutia on Cuba as well as to the decline of the Cuban solenodon and the native rabbit on the Japanese island of Amami-Oshima.

Similarly, introduced grazers have devastated islands. For instance, on the large subantarctic island of South Georgia, over 1,000 miles east of South America, 17 reindeer were introduced by Norwegian whalers, probably for fresh meat supply, in the early 20th century. In 50 years, the population increased to over 3,000, and they have heavily damaged the large, prominent native lichens and tussock grassland. The vegetation change has in turn harmed seabird populations. As another example, European rabbits, introduced to at least 800 islands worldwide, have devastated vegetation on many of them. On Australia, the largest of these islands, their biggest ecological impact is stripping and killing seedling trees and perennial shrubs, especially of acacias, thereby contributing substantially to deforestation and desertification. Rabbits are also present on 48 islands off the Australian coast, on many of which they have completely destroyed or transformed the native vegetation, often leading to abandonment by seabirds. Through competition for burrows or food, rabbits are believed to have caused the extinction of two burrowing marsupials, and by destroying pasture they have constricted the range of the southern hairy-nosed wombat. On tiny Round Island, off the coast of Mauritius in the Indian Ocean, rabbits devastated native vegetation and even threatened an endemic palm species with extinction until a successful eradication program was completed in 1986.

2.6 How are introduced species distributed among habitats, and how do they get around?

Disturbed habitats are particularly susceptible to invasion. Consider roadsides, railroad rights-of-way, abandoned fields, and similar places: such habitats are far more dominated by introduced plants than are pristine habitats. Ecologist Mark Davis and colleagues have suggested that perturbations of various sorts liberate resources, such as soil nutrients or light, and thereby allow introduced species to gain a foothold. If there were no fluctuations in resources, the resident species would have evolved adaptations allowing them to monopolize resources so efficiently that few would be available for a potential new colonist. However, a sudden disturbance can generate a surge of some resource that cannot be immediately monopolized by native residents, providing the opportunity for invasion.

Many disturbed habitats produced by human activities also constitute new habitats for a region, such as golf courses in desert areas and agricultural fields in cleared forests. These are unlike anything previously found in these areas, so it is not surprising that, whatever the regional native species are, they do not thrive in them. Rather, the species that do well in such habitats are those previously adapted to that particular kind of habitat. Thus, for example, many weeds of agriculture have become cosmopolitan. Once their seeds were spread inadvertently by humans (for example, as seed contaminants), the agricultural habitat that they found in their new homes was quite similar on all continents. For instance, wheat fields in Europe, Asia, North America, and Australia are much more similar to one another than are forests of the same continents.

It is therefore not only disturbance that is important in facilitating invasions but also the type of disturbance and the extent to which it produces a novel habitat. The proof of this assertion can be found by looking at naturally chronically disturbed systems—places like high-energy beaches or plant communities subjected to frequent naturally originating fires. None of these ecosystems are characterized by heavy invasion. In fact, most of them have proportionally fewer invaders than pristine habitats that are not routinely disturbed, probably because few introduced species are adapted to the harsh circumstances generated by the particular sort of chronic disturbance. A good example is the longleaf pine–wiregrass community that formerly dominated over 50 million acres of the American southeastern lowland. This is what is known as a *fire disclimax* community because it is maintained by frequent lightning-caused fires, which prevent the growth of hardwood trees that would otherwise replace the pines. Longleaf forest is now greatly reduced, so that original examples of such forest total at most a few thousand acres. However, these forests are scarcely invaded, because the many nonnative plant species that are common nowadays within this region, such as ornamentals and various grasses, are unable to tolerate the fire regime. Unfortunately, this situation may be changing, as fire-tolerant Asian cogongrass spreads in the South.

Plant ecologists Andrew S. MacDougall and Roy Turkington, observing the frequent association of invasions with habitat disturbance, have proposed the "passenger–driver model" of invasion. Using a car analogy, they suggest that invasive plants are more frequently the "passengers" of some habitat disturbance than "drivers" of habitat change. In this view, attempting to prevent or manage invasions, as described in Chapters 9 and 10, is an inefficient approach to the problem, and we would do far better instead to ameliorate whatever other environmental change is truly driving the invasion. However, given the increasing number of observed impacts of introduced plants on entire ecosystems (Chapter 3.1), plus a tally by ecologist Patrick Martin and colleagues of many undisturbed plant communities that have nevertheless been substantially invaded, it is questionable whether the passenger–driver model is really an appropriate guide for managing many plant invasions. It is also now apparent that some species that begin as passengers end up as drivers of habitat change. For instance, New Zealand pygmyweed, an aquatic plant that has invaded Great Britain, continental Europe, and part of the United States, appears in Great Britain to be a passenger of other environmental changes in some places, such as reed beds, where it is facilitated by nonnative sika deer, which enhance light by eating native reeds more than pygmy-weed and by creating paths that spread the invader. In other habitats, pygmyweed is a driver of habitat change, as on the edges of temporary ponds, where it outcompetes native species. There may be a temporal shift from passenger to driver. Consider also a variety of the marine alga sea grape that has invaded the Mediterranean Sea (Chapter 4.4); it appears to begin as a passenger as it invades reefs in human-disturbed areas, but after it invades it accumulates sediment that favors growth of an algal turf that then transforms the entire reef.

Freshwater habitats, even undisturbed ones, are particularly afflicted with biological invasions. In some ways, lakes and ponds are like islands—they are aquatic habitats (often small ones) isolated from one another by great expanses of terrestrial habitat that cannot easily be traversed by aquatic species. Thus, like islands, many of them have relatively few native species and lack certain entire groups of species found in larger bodies of water. Many high-elevation lakes are naturally fishless, for example, and introduction of various fish species (especially trout) into many of these lakes has led to enormous impacts on the resident community, just as the introduction of mammalian predators and grazers has devastated island communities. Similarly, introduction of crayfish into lakes previously lacking them has had drastic impacts on the entire community. For instance, the arrival in 1997 of the American red swamp crayfish in Lake Chozas, Spain, led to the elimination of almost all submerged vegetation and transformed the water from clear to turbid. Many aquatic invertebrates subsequently disappeared from the lake, and its use by waterfowl plummeted.

An additional reason for the high rate of invasion impacts in freshwater habitats is heavy use of water by humans for transportation, commerce, and recreation. This water usage combines with the many dispersal mechanisms of aquatic species to introduce plants and animals to distant sites and also to allow them to spread remarkably rapidly once they have established populations in their new homes. For example, Eurasian zebra mussels, once they reached North America in ballast water or perhaps by fouling a boat hull, spread quickly downstream through the eastern United States and Canada by virtue of their planktonic larvae or by adults or larvae attaching to plants that subsequently broke off and drifted. They were moved upstream and into new drainages by fishermen, who transported them inadvertently—the larvae in the water of live wells and adults and larvae encrusted on hulls, anchors, and waterweeds attached to the latter. For introduced fishes, in addition to their own powers of locomotion, deliberate movement between drainages by sport fishermen seeking new opportunities is common, while bait fish may be deliberately or inadvertently released great distances from their origin. Many aquatic plants and animals also produce enormous numbers of offspring that easily float downstream within drainages and can be transported by humans just as zebra mussels have been moved about.

Construction of canals can lead to a wave of invasions by drift, active movement, or incidental transport by ships. The best-known example is the mass migration of species from the Red Sea to the Mediterranean Sea after the construction of the Suez Canal. These species are termed *Lessepsian migrants* after Ferdinand de Lesseps, who headed construction of the canal, and they include fishes, worms, crustaceans, and sea slugs—over 300 species total. Strikingly, very few species have invaded in the opposite direction. Once present in the eastern Mediterranean, Lessepsian migrants can spread very quickly westward. An extreme example is the bluespotted cornetfish. This large, striking predator took over a century to traverse the Suez Canal. Once it reached the Mediterranean in 2000, however, it took only two years to spread westward to Italy and by 2010 had dispersed all the way to Spain (Figure 2.2). Even more remarkably, this initial invasion probably consisted of only two females. The spread of warmwater Lessepsian

Figure 2.2 Bluespotted cornetfish, predator that spread throughout the Mediterranean Sea beginning in 2000. (Drawing courtesy Ernesto Azzurro.)

migrants toward the northern and western Mediterranean is probably aided nowadays by warming of the Mediterranean.

The European canal network connecting the North, Baltic, Mediterranean, White, Black, Azov, and Caspian Seas has 30 main canals and stretches over 17,000 miles. At least 65 species are believed to have been dispersed between drainages through this network, including high-impact invaders such as the zebra mussel and Asiatic clam. For instance, a Caspian shrimp known as the bloody red mysid reached both the Baltic and the North Sea by this route. The growing recognition of the threat posed by invasions has recently led to the occasional implementation of mechanisms to hinder such movement. For example, an electric barrier was installed in the Chicago Sanitary and Ship Canal to prevent Asian silver carp and bighead carp, which had already invaded the Mississippi River drainage, from reaching the Great Lakes. This barrier has failed; carp DNA has recently been found in the Great Lakes (Chapter 12.6). In any event, such a barrier would not have prevented movement of fouling organisms or those in ballast water. Probably the most famous invader of the Great Lakes via canals is the alewife, an Atlantic fish that reached the Great Lakes through the Welland Canal, with enormous subsequent impacts on the food web.

In sum, any habitat can be invaded if the right species is introduced, but habitats created by human activities are particularly prone to invasions. This is largely because native species have not evolved adaptations to such habitats. Islands, especially small, remote ones, are also particularly likely to be damaged by invasions, both by virtue of their small size and their frequent lack of certain sorts of native species (Chapter 2.4). Aquatic habitats are frequently invaded, at least partly because of the myriad ways aquatic species can move about, sometimes en masse, as through canals or in ballast water. Other habitats, such as boreal forests and most tropical forests, are so far relatively infrequently invaded. Whether this is due to a stroke of luck or because of some inherent features that confer resistance is unknown, but the continuing flow of propagules of nonnative species into such habitats may increase the degree of invasion (Chapter 2.3).

3

ECOLOGICAL EFFECTS OF INTRODUCED SPECIES— STRAIGHTFORWARD IMPACTS

Invasive nonnative species have a huge number of ecological impacts. Some are apparent to the most casual observer; others are less obvious, and some, while important, are so subtle that they cannot be detected without intensive research. However, just because they are subtle need not mean they are unimportant. Many impacts are simply bizarre; who could have imagined that invasion by hydrilla, an Old World aquatic plant, could lead to the death of bald eagles in Georgia? The birds had eaten coots that had consumed neurotoxin-containing algae that grew on the hydrilla mats! Or who would have guessed that an introduced tropical American weed, bitterbush, could threaten the future of Nile crocodiles in a lake in Africa by shading nesting sites, thereby lowering the temperature of developing eggs so that no males are produced (crocodile sex is determined by egg temperature)? Many impacts of invasions are like these—so idiosyncratic that no one could have predicted them. Often one invader will have several impacts on native species and ecosystems, and sometimes activities of different invaders will combine to exacerbate the total impact beyond what might have been expected from the individual impacts of each invader.

No two invasions are identical, but most impacts fall into several well-defined categories. Some of these categories describe impacts that usually affect one or a few populations, while others commonly affect many species at once, even entire ecosystems.

3.1 How do biological invasions modify habitats?

Because so many species are closely tied to particular habitats, impacts that greatly change the habitat can ripple through an entire community. For example, in the 18th and 19th centuries, the northeastern North American coast was composed of extensive mud flats and salt marshes. Nowadays, it is usually characterized by rocky beaches. This dramatic change is all due to the European common periwinkle snail (see Chapter 11.3), introduced (probably for food) to Nova Scotia around 1840. It slowly spread southward, eating algae on rocks and also rootstocks of marsh grasses and transforming vegetated coasts into barren rocky shores. Thus, the periwinkle has modified the entire physical structure of the intertidal zone, and in the process it has affected many other species. It displaces native snails and prevents barnacle larvae and young seaweeds from settling, and marshland birds move away. In the eastern Mediterranean Sea, two Red Sea rabbitfishes that arrived via the Suez Canal similarly change the habitat by grazing algae. They create large areas of bare rock known as *barrens* that are characterized by low biodiversity, largely through loss of the habitat for marine invertebrates that live in the algal canopy and that are the main food items for carnivorous fishes of the region (Figure 3.1).

By contrast to the periwinkle and the rabbitfishes, which remove the original structured habitat, the zebra mussel (native to southern Russia) has greatly modified the habitat of many ecosystems by creating a new habitat. It was probably

Figure 3.1 Rabbitfish from Red Sea creating a barren area on floor of Mediterranean Sea. (Photograph courtesy Zafer Kizilkaya.)

transported to North America in ballast water or attached to a ship's hull, and by 2000 it had spread over much of the eastern United States and Canada. Most public attention has been focused on its economic impacts through fouling and clogging water pipes, with costs to date estimated at billions of dollars. But water pipes aside, its ecological impacts are equally drastic. Settling in dense aggregations that smother native mussels, it has converted the substrate in some areas into a jagged mass of mussel shells. In addition, zebra mussels filter water at a prodigious rate, thereby increasing water clarity, decreasing phytoplankton densities, and affecting populations of fish, zooplankton, and other invertebrates that feed on plankton. The very existence of many native mollusk species is threatened.

Although introduced animals can obviously wreak significant havoc, introduced plants are the more frequent causes of massive habitat change, simply because plants often constitute the habitat for an entire community and because terrestrial, aquatic, and marine plants can overgrow large areas. A notable example is Brazilian pepper in south Florida, a small tree introduced in the 19th century that spread to dominate over half a million acres by 1993, often as a monoculture. Birds and insects that inhabit Brazilian pepper forests differ greatly from those in the communities these forests have replaced, such as pinelands and mangrove swamps. In addition, Brazilian pepper has subtler impacts that nonetheless affect entire ecosystems. For example, where Brazilian pepper invades pineland savannas, it tends to decrease fire frequency and temperature, thereby harming the native plants that depend on occasional fires. Brazilian pepper also appears to be allelopathic toward some native plants—it produces chemicals that inhibit their growth. Another prominent example is that of rock snot, a Northern Hemisphere diatom (single-celled alga) introduced to New Zealand, which produces thick, gooey mats that blanket streambeds, affecting both resident insects and the fish that feed on them as well as producing treacherous footing for fishermen. Rock snot has recently invaded areas of North America far south of its original range, perhaps inadvertently carried on fishing boots or gear.

Plants can change entire ecosystems even without overgrowing them by modifying various ecosystem features and processes. For example, in south Florida, Australian paperbark trees have traits (including spongy outer bark and highly flammable leaves and litter) that lead to increased fire intensity and frequency. These changes, in turn, have helped paperbark replace native plants not adapted to this fire regime. About half a million acres of sawgrass and muhly grass meadows were gradually transformed into paperbark forests. A further threat has now arrived in the region—Old World climbing fern. This plant climbs up paperbark and what native trees are left, and it thus transmits fire to the canopy of trees that might otherwise have survived a ground fire. This is one among many cases in which introduced plants affect entire ecosystems by modifying natural fire regimes. For instance, cheatgrass has replaced native plants on hundreds of thousands of acres in the American West, largely by fostering and tolerating fires (Chapter 1.3). Similarly, in Hawaii, larger and more frequent fires in the wake of invasion by South American tufted beardgrass and African molasses grass have converted high elevation woodland to grassland.

Another way introduced plants can affect entire ecosystems is by changing hydrology. In the American Southwest, Mediterranean salt cedars can cause severe water loss in arid areas because of their deep roots and rapid transpiration. In California, salt cedar drained the surface water of a large marsh (Chapter 1.3). Similarly, in South Africa, invasion by species of Australian *Acacia* and other trees has decreased net water runoff in large ecosystems by 6%.

Introduced plants can also modify an ecosystem's nutrients. On the volcanic island of Hawaii, the Atlantic nitrogen-fixing firebush has invaded geologically young, nitrogen-poor areas. Because of a lack of native nitrogen-fixing plants in mid- and high-elevation areas, the native plant species have evolved adaptations to the nitrogen-poor soil, while many introduced species, not having evolved in such soil, are poorly adapted to it. Ecologist Peter Vitousek and colleagues have shown that firebush is gradually replacing the native dominant tree species on much of the island and is increasing the concentration of nitrogen-containing compounds in the soil. Now there is the daunting prospect that a wave of plant invaders will be able to use the increased nitrogen to establish over large areas. Nitrogen and water content of the canopy have already doubled where firebush has replaced native forest. Elsewhere, invasion by other nitrogen-fixing plants, such as yellow bush lupine in North America and coojong (a type of acacia) in South Africa, have had similar impacts. Other plants have ways of increasing soil concentrations of phosphorus, the other key limiting plant nutrient. For instance, bridal creeper from southern Africa has invaded southern Australia with subsequent replacement of some native plant species by introduced species. Peter J. Turner and colleagues have shown how the physiology and morphology of bridal creeper have adaptations that permit soil around this plant to accumulate increased phosphorus in phosphorus-poor soils. As the low levels of phosphorus had been preventing other introduced plants from establishing, bridal creeper is changing the rules of the game, so to speak, and is precipitating other invasions. Other introduced plants bring about the same increase in soil phosphorus in Western Australia, with the same general result.

Introduced species of the pine family from the Northern Hemisphere, such as Monterey pine and Douglas fir, have widely invaded in the Southern Hemisphere, affecting South Africa, South America, Australia, and New Zealand. Among other impacts, they usually produce a litter of much poorer quality than that of the native vegetation—more acid and with higher levels of defensive chemicals. This change impedes decomposition and also reduces populations of many soil animals. Introduced animals can also greatly affect nutrients and, through the impact on nutrients, many soil organisms. For instance, red deer introduced to New Zealand tend to eat forest understory plants that produce high-quality litter, and these plants tend to be replaced by plants that are avoided by the deer and that produce lower-quality litter. Soil animals are thus greatly affected, as are the carbon and nitrogen cycles.

Introduced earthworms greatly change the litter and soil habitat, but the impact of these changes on the native species has not been adequately understood because these effects occur underground, out of sight where they are difficult to study. All earthworms in much of Canada and the northern United States are Eurasian immigrants that arrived as glaciers retreated a few thousand years ago. Research by Cynthia Hale and her colleagues in previous worm-free areas in Minnesota and Wisconsin shows that worms remove much of the litter layer, thus greatly reducing success of tree and herb seedlings and also leading to loss of ground-nesting birds such as the ovenbird. The introduced earthworm *Pontoscolex corethrurus*, in parts of South America and Australia, eliminates native worms and greatly compacts the soil, affecting the entire soil habitat (Chapter 2.3).

Introduced North American beavers also create entirely new habitat. They are known as *ecosystem engineers* because their activities fundamentally change the physical structure of the environment. Approximately 25 pairs were introduced in 1946 to the main island of Tierra del Fuego, at the southern tip of South America, in an attempt to start a fur industry. Now at least 50,000 individuals inhabit the area, and they have spread to other islands, causing major ecosystem changes by building dams and demolishing trees (Chapter 1.3).

3.2 How do invaders compete with native species?

Competition between species can take two forms. In one form, called *interference competition*, individuals of one species prevent individuals of the other from garnering resources by fighting, for example, or intimidation. In the second form, competition can come about simply because two species, using the same resource, lower its supply, thereby making it more difficult for individuals to satisfy their needs. In this latter type of competition, termed *resource competition*, two species can affect one another even if individuals never come into contact (as when diurnal and nocturnal species compete for the same food). In either case, a key requirement for two species to compete is the presence of a population impact on at least one of the species. That is, the size of the population of at least one species must be lower than it would have been without the competition. If we simply observe an individual, or many individuals, of one species eating food that individuals of another species would eat or chasing away individuals of the other species, we cannot be certain competition is occurring, even though we may strongly suspect that it is.

Interference competition has been documented frequently between introduced and native species, and it is assumed to occur even more frequently though the population impacts have not been proven yet. Several introduced ant species affect native ants by attacking (and often killing) them, especially at food sources. In North America, densities of native ant species plummet in the vicinity of colonies of both the red imported fire ant and the Argentine ant. Elsewhere, similar competitive impacts on other ants have been documented for invasions by the yellow crazy ant, the big-headed ant, and the little fire ant. In Australia, European red foxes threaten rare native quolls by driving them from food sources.

Plants, although they do not "behave" and attack as animals do, have traits that are analogous. For example, the African crystalline ice plant has invaded coastal grassland in California. It accumulates salt, which remains in the soil when the plant decomposes, thereby excluding native California plants that cannot tolerate salt. A related phenomenon, allelopathy, has been cited for several introduced plants, although each case has been controversial. Allelopathy is defined as the secretion of a chemical by one plant species that suppresses germination or growth of another species. It is admittedly difficult to conduct sufficiently controlled experiments to demonstrate the precise impacts of minute amounts of potentially allelopathic chemicals. However, research by ecologist Ragan Callaway and his colleagues strongly suggests that two of the most aggressively spreading weeds in the American West—spotted knapweed and diffuse knapweed—inhibit native plants by allelopathic root secretions.

In other instances, introduced species have largely replaced native species by resource competition. For example, as cited earlier, in Great Britain the introduced eastern gray squirrel of North America has caused populations of the native red squirrel to decline by outcompeting the latter for nuts. The invading squirrel forages more efficiently for food, particularly as deciduous forests gradually replace pine forests in Great Britain. An eradication campaign in Italy against a small spreading population of the gray squirrel was halted by an animal-rights lawsuit, and there is now concern that the same replacement may occur in mainland Europe. Similarly, in Australia, the European rabbit outcompetes native marsupials such as wallabies and rat kangaroos for food, contributing to their decline. Finally, the Asian house gecko, which is bigger and faster than native geckos, outcompetes the natives for insect prey on Pacific Islands, particularly around building structures and other human habitats (Chapter 1.3).

A famous aquatic case of resource competition involves the alewife, an Atlantic coastal fish introduced to the North American Great Lakes by construction of the Welland Canal (Chapter 2.3). Through competition for zooplankton populations fed on by native species, the alewife contributed to the disappearance of several large native fishes. In the Baltic Sea, introduction of a comb jellyfish, the sea walnut, from the western Atlantic Sea led to drastic declines of several native fishes, which were outcompeted for food—zooplankton and fish eggs and larvae. For plants, several cases of root competition for resources such as minerals and water have been demonstrated experimentally. For instance, laboratory and field experiments in Maryland showed that Japanese honeysuckle outcompeted native sweet gum saplings by root competition, in addition to possible competition for light among the leaves. Similarly, in California, soft brome, an introduced annual grass, outcompeted blue oak seedlings for water and nitrogen. In the American Midwest, introduced purple loosestrife outcompetes native winged loosestrife for visits by pollinating insects.

Invaders that compete with natives need not be closely related to them (Chapter 1.3). The competition in New Zealand between an introduced wasp and native birds for insect "honeydew" is a prime example. In another case, the Mozambique tilapia has been introduced to many parts of the world as a fish for sport and food. It is believed to have caused the extinction of an endemic duck on a Pacific island, the Rennell Island teal, by outcompeting it for food, and it is similarly suspected of threatening the white-headed duck in Europe.

3.3 What are the impacts of introduced predators?

Many introduced species prey on native species, sometimes driving them to local extirpation or global extinction. Predation by the introduced sea lamprey, in combination with other factors, led directly to the extinction of three endemic Great Lakes fishes: the longjaw cisco, deepwater cisco, and blackfin cisco. Along with overfishing, watershed deforestation, and pollution, lampreys devastated populations of all large native fish, even when they did not cause extinction. Economic repercussions were dramatic, as catches of many species fell by 90% or more. Declines of these large fish rippled through the food web, with increases in populations of several smaller fish species that had been their prey. In a chain reaction, these newly inflated populations of small fish then caused decreases in populations of still smaller animals that were their prey.

The many extinctions resulting from the introduction of the Nile perch into Lake Victoria and the introduction of rats on islands worldwide (Chapter 1.3) are but two examples of huge impacts of introduced predators. Particularly tragic are the consequences of several deliberate introductions of predators for biological control of previously introduced pests. Extinctions caused by the introductions to islands of the rosy wolf snail in failed attempts to control the giant African snail and of the small Indian mongoose in failed attempts to control rats (Chapter 2.5) are among them (Figure 3.2). In spite of ornithologists' concerns, ferrets were introduced to New Zealand in 1879 and stoats and weasels in 1884 to control introduced rabbits. All three species had little effect on rabbits; however, they prey on many ground-nesting birds, and there is good evidence for a population effect of the stoats in combination with predation by ship rats and Pacific rats on some of these unfortunate prey species. In a last example, the multicolored lady beetle was introduced from Asia to the United States to control aphid pests of agriculture, but it has reduced populations of several native lady beetles as well as another introduced lady beetle by preying on them.

Figure 3.2 North American rosy wolf snail (R) attacking giant African snail (L), both introduced to Hawaii. (Photograph courtesy Kenneth Hayes.)

Feral cats have had disastrous impacts on islands worldwide. Consider Ascension Island in the central South Atlantic, one of the most important warm water seabird stations in the world, with breeding populations of 11 species. Primarily because of cats, only one species, the sooty tern, now breeds on the 34 square mile main island. The other species are restricted to tiny Boatswain's Island, approximately 0.04 square miles. In addition, cats have extinguished two endemic Ascension landbirds, a rail and a night heron. A similar situation exists in New Zealand and elsewhere, where some birds persist only on tiny offshore islands that introduced predators have not yet reached. Fortunately, this situation is changing as technology to eliminate such introduced predators improves (Chapter 9.3).

Rats are also disastrous introduced predators. Of about 50 rat species, 4 readily become feral—the Norway rat, the ship or black rat, the Oriental house rat, and the Pacific rat. All have been very destructive, especially the ship rat because it is a skillful climber. Ship rats extinguished the populations of all three endemic New Zealand bats in just one year on Big

South Cape Island (2,300 acres) off Stewart Island, at the same time eliminating five bird species there. One of the three bats is probably now globally extinct, and another is endangered. Rats have also been a disaster for birds on isolated oceanic islands worldwide. The only islands on which introduced rats have not greatly affected birds are those that had endemic rats to begin with or those with land crabs, probably because bird species on such islands evolved to avoid predators. Land crabs are widespread on tropical islands—for example, the famous red land crab on Christmas Island in the Indian Ocean. Endemic island rat species were less numerous, and populations of several of these native rats are now dwindling because of predation by introduced rats or competition with them.

The introduced sea walnut has had not only a competitive impact on Black Sea fishes but also enormous predatory effects. At one point, the biomass of the sea walnut population was 10 times greater than that of all other planktonic species combined. Able to eat animals up to 1 cm long, it devastated prey and competitor species until its own population declined because of food shortage. However, it was still common until the late 1990s, when another comb jellyfish (*Beroe ovata*) arrived in the Black Sea and turned out to be a voracious predator on the sea walnut, drastically reducing its population.

The brown tree snake, introduced in cargo from the Admiralty Islands (Chapter 1.3), has eliminated 15 native forest bird species on Guam, leaving but one species in forests that are now eerily silent. Another snake invader, the Burmese python, is perhaps best known for a photograph showing the outcome of a deadly struggle between a python and a native alligator in Florida. However, even more chilling than the photograph is the fact that in Everglades National Park several mammal populations have plummeted dramatically since the proliferation of the python, as evidenced by extensive road surveys. Observations of raccoons declined by 99.3%, opossums by 98.9%, and bobcats by 87.5%. Rabbits, red foxes, and gray foxes ceased to be recorded entirely.

Like some introduced plants, invasive predators can generate effects that ripple through the entire ecosystem, affecting everything from animals to plants to soil organisms. For instance, Norway rats and ship rats on offshore islands of New Zealand prey on petrels, shearwaters, and other seabirds. These birds feed at sea but nest in burrows on the islands, often in great densities. They transport nutrients from the sea to the land in the form of guano, feathers, carcasses, eggs, and food they bring their chicks. Predation by rats interrupts this flow, resulting in cascading impacts on nutrient cycles, plants, and belowground organisms. Introduced Arctic foxes preying on seabirds in the Aleutian Islands generate a similar ecosystem-wide impact. The consequent reduced transport of nutrients from sea to land decreases soil fertility, transforming grasslands into dwarf shrub- and forb-dominated vegetation. In Great Britain, predation of earthworms by the New Zealand flatworm (Chapter 2.4) changes soil characteristics such as porosity and drainage, thereby affecting aboveground plants and animals.

Many introduced predators that have had enormous detrimental impacts, such as the mongoose, the rosy wolf snail, and the stoat, were introduced for biological control of previously introduced pests. The cane toad makes for a superb fourth example. Introduced from the New World to Australia in a futile attempt to control two previously introduced beetle pests of sugar cane, it multiplied rapidly and spread widely. Although the toad does prey on many species, it is not known to have substantially impacted its prey. Surprisingly, its main conservation impact has been that native predators attacking it, such as quolls, goannas, death adders, and red-bellied blacksnakes, often die because of toxins produced by the toad. Although the full effect on populations of these toad predators is unknown, some of these species are rare, and it is suspected that toad-induced mortality is important. The general problem with these four biological control disasters has been that all involved generalized predators that consume a wide variety of prey items. It was therefore almost inevitable that, whatever they did in terms of the target pests (in all four cases, they appear to have failed to control them), they were likely to cause problems for native prey.

However, this is not to say that there have been no instances of successful biological control by predators without known nontarget impacts (Chapter 10.3). For example, in the 1970s or 1980s the South and Central American orthezia scale insect reached the Atlantic island of St. Helena and threatened the very existence of gumwood, the endemic national tree of the island. By 1993 the predatory South American ladybird beetle *Hyperaspis pantherina* was introduced, and it quickly brought down the orthezia scale population. Another successful example is of an introduced predator on an introduced herbivore on introduced trees. Britain has no native spruce but does have extensive plantations of Norway spruce and Sitka spruce. In 1982, the great spruce bark beetle reached Great Britain from continental Europe and became a major spruce pest. A predatory beetle, *Rhizophagus grandis*, was brought from Europe and has controlled the bark beetle well in a number of locations.

3.4 What are the impacts of introduced herbivores?

Introduced herbivores, if they affect a dominant plant, can change the entire physical structure of an ecosystem, often with follow-on effects that can harm many resident species. The impacts of invading rabbits and reindeer on islands (Chapter 2.5) are often in this category. For instance, introduced rabbits on subantarctic Kerguelen Island have greatly reduced the population of the endemic Kerguelen cabbage, threatening the very existence of this large, dominant cover plant and also forcing a native wingless fly that specializes on this cabbage to use decaying seaweeds, a habitat to which it is ill-adapted. Similarly, the introduction of beavers to Tierra del Fuego, described already, has many consequences related to its conversion of forests to meadows. For instance, many introduced ground cover plant species that are rare or absent in the forest have colonized the meadows created by beavers. Camels, introduced to Australia in 1840 because they were better suited than horses to the dry climate, were released by the thousands in the 1920s and 1930s as motor vehicles became available and camels were no longer needed to haul freight. Today, over a million wild camels damage trees and shrubs by browsing and trampling on ground cover plants and shrub seedlings. Camels severely defoliate trees, shrubs, and vines of preferred species, and their browsing on young plants inhibits flowering and fruit production. They can even locally extirpate preferred species.

Several invasive plant-eating insects have wrought havoc in North America. The European gypsy moth (Chapter 1.3) occupies over 200,000 square miles of the northeastern United States and eastern Canada. The moth feeds on many woody plants, especially oaks and quaking aspen. An Asian strain of the same species has appeared in Oregon and British Columbia. Defoliation by the gypsy moth weakens trees, increasing their susceptibility to other insects and diseases. In some areas, repeated defoliation has caused up to 90% mortality of preferred host trees, thus greatly changing forest composition. A major gypsy moth infestation has many other impacts. Litter amounts and decomposition increase, augmenting nitrogen loss in stream flow, while both defoliation and reduction of oak mast production affect bird populations. In the highest elevations of the southern Appalachians, an Asian aphid called the balsam woolly adelgid has killed most individuals of the previously dominant tree, the Fraser fir. A relative of this species, the Asian hemlock woolly adelgid, has killed most hemlock trees in extensive forests of the Northeast and is now killing hemlocks in the southern Appalachians, where hemlock dominates glades that constitute distinct habitats along water courses. The destruction of hemlocks by the hemlock woolly adelgid changes both water and nutrient flows. Aphids and adelgids are in the order of true bugs, Homoptera, which contains many extremely damaging insect invaders, including scale insects and whiteflies. In Florida, a 2009 homopteran arrival from Central America, the gumbo-limbo spiraling whitefly, has been devastating gumbo-limbo trees. This tree species is native and prominent in south Florida forests and is also used widely as an ornamental plant. The same whitefly also attacks a wide variety of other trees, such as oaks and palms.

Perhaps the best known population impact of herbivores is economic damage caused by insect pests attacking agricultural crops. A striking example is the Russian wheat aphid, a native of southeastern Europe and southwestern Asia that reached Mexico in the 1980s, arrived in the United States from Mexico in 1986, and quickly spread through the western part of the United States and Alberta and Saskatchewan. It attacks not only wheat but also barley and, less intensively, rye and triticale. It has cost over $1 billion so far in yield losses and control costs, and it has led to the near elimination of wheat and barley crops in some regions. In addition to harming crops, it has ecological impacts: it infests crested wheatgrass, extensively planted for soil conservation, and has spurred the introduction of the Eurasian seven-spot lady beetle, widely distributed to combat this aphid and responsible for displacing native lady beetles in several areas.

Some introduced herbivores affect both agricultural crops and native vegetation. For instance, the pink hibiscus mealybug, native to Asia and Australia, reached the United States near the turn of the 21st century and attacks well over 100 plant species, including vegetables and ornamental plants as well as some forest trees among other native plants. A recent invader in the southern United States, the Asian redbay ambrosia beetle, first attacked native redbay laurel trees, infecting them with the laurel wilt fungus, a pathogen that probably arrived with the beetle. Recently both the beetle and fungus have spread to avocados (a member of the laurel family) in Florida. Worldwide, perhaps the worst introduced pest of agriculture is the sweet potato whitefly, believed to have originated in India but now found on all continents except Antarctica and on many islands as well. It attacks over 900 plant species, including many crops, but also native plants in many locations, and furthermore transmits highly damaging plant viruses. The tremendous breadth of host plants has led scientists to study this species intensively, and it is now thought to be not just a single species but a complex of closely related, nearly indistinguishable species with various host preferences.

3.5 What roles do introduced parasites and diseases play?

Introduced parasites and pathogens routinely devastate native populations. If the affected native species is a key part of an ecosystem, its decline can precipitate a cascade of effects that substantially transform the entire ecosystem. The fungus causing chestnut blight, discussed previously, is a prime example. Fungus-like pathogens called water molds have invaded many forests worldwide, sometimes devastating entire ecosystems. For instance, the water mold jarrah dieback has killed millions of native trees of many species in Australia, damaging forests so much that rare native plants and marsupials are endangered. The geographic origin of jarrah dieback is still uncertain, but it has spread worldwide. A related water mold, sudden oak death, struck California in 1995. Nearly all trees in some California forests proved susceptible to this pathogen, including redwoods and Douglas firs, although most survive. Oaks, however, have suffered high rates of mortality. The ultimate impact of sudden oak death on forest ecosystems previously dominated by oaks depends on which species replace them, which will unfold over many years. In the meantime, this same water mold has reached Europe with amazing speed, although molecular evidence suggests that the European invasion originated from a geographical source (as yet unidentified) different from the California invasion.

A number of other water molds in the same genus as jarrah dieback and sudden oak death have recently invaded elsewhere, attacking various native trees. Two examples are the *Phytophthora* disease on alder trees in Europe and a needle blight on introduced Monterey pines in Chile. This sudden advent of water mold invasions is spurred by two forces. First is the proliferation of global trade, which carries the spores as hitchhikers. Second is the fact that members of the genus containing these pathogens readily hybridize, generating new combinations of genes. Some of these new combinations are highly pathogenic.

Pathogens that attack plants rather than animals are more likely to affect entire ecosystems when they invade, because plants dominate the structure and function of most ecosystems. However, at least one invasion by an animal pathogen—rinderpest—has been equally devastating at a regional scale. Rinderpest is a viral disease of ungulates, closely related to measles and distemper, and was introduced to southern Africa from Arabia or India in cattle in the 1890s. The spectacular pandemic caused massive mortality throughout sub-Saharan Africa of many native grazing animals such as elands, giraffes, impalas, and warthogs. Wildebeest and buffalo populations, particularly hard hit, plummeted 95% in just two years. The geographic range of some of these species in Africa still reflects rinderpest mortality, even though the disease was finally eradicated in 2011 (Chapter 9.3). When the ungulate populations crashed, carnivore populations crashed also, and starving lions wandered into villages in search of food. People abandoned large areas because their livestock all died, and, with the population crash of the big herbivores and the departure of people, the landscape of places like the Serengeti Plains was transformed from an open grassland to a densely wooded, nearly impenetrable savanna. Several wildlife populations, especially that of wildebeest, began to recover over the past two decades as the eradication program succeeded, setting in motion several linked changes in the plains ecosystem.

Although not all parasite and pathogen invasions affect whole ecosystems as in the previous examples, many devastate populations of single species or groups of species. A well-known plant pathogen of this is sort is the fungus (recently shown to consist of two separate species) that causes Dutch elm disease. The name is a misnomer, as the fungus is probably native to the Himalayas. In the late 19th or early 20th century it was introduced into Europe from the Dutch East Indies, and a Dutch biologist first described the pathogen—hence the name. Dutch elm disease spread through continental Europe, ravaging elm populations. The first known invasion into the United States was in Cleveland in 1930, after which the disease spread rapidly in natural and urban elm populations throughout North America. A second invasion into North America occurred in 1945, starting in Sorel, Quebec. By the 1950s and 1960s, most elms in North America had died. Elm was a common urban tree; many towns had lined their main streets with it, leaving a legacy today of hundreds of elmless Elm Streets. Elm was never, at least in historic times, a dominant forest tree as chestnut and some oaks were, but Dutch elm disease was noticed more than any other plant pathogen in the United States because of the urban prominence of elms.

The fact that elm was so dominant in cities, towns, and villages, with many elms planted together, likely contributed to the rapid spread of the disease. Two species of bark beetles, one native and one introduced from Europe, are vectors of the fungus. But the fungus can also spread underground, through roots, when trees are planted so close together that roots are intertwined.

Long after the initial invasion, Dutch elm disease declined in Europe, though not in the United States. However, it reemerged in Europe in an even more virulent form that was introduced from North America on diseased timber, first striking in Great Britain in the 1960s and then spreading to most of Europe. Molecular differences from the older strain led specialists to classify the newer one as a new species, and it has been replacing the older strain on both continents. This new, virulent species may have arisen as a hybrid between the older species and another fungus that was introduced to North America.

Native birds of Hawaii have been devastated by two pathogens—avian malaria (a protozoan) and avian pox (a virus)—that were inadvertently introduced with Asian songbirds in the first half of the 20th century The introduced birds were themselves resistant to these diseases, having coevolved with them for millennia, but the native Hawaiian birds were devastated, particularly by avian malaria. Avian malaria was transmitted to the native birds by introduced mosquitoes, especially the American southern house mosquito, which had been introduced a century earlier; mortality rates were 65–90% after a single bite. Native Hawaiian birds are highly threatened (Chapter 2.5): 14 of the 38 native Hawaiian forest birds that existed when European colonization began are already extinct, and 15 are on the federal endangered species list. The main reason is surely habitat destruction (about a third of the native forest remains, and much less on some islands), but avian malaria is a huge contributing factor. The chief problem here is that most of the remaining native forest, which is the normal habitat of these birds, is in high elevations, but most of these birds are not well adapted to high elevations. Substantial areas of native forest remain in the lowland areas, but native birds cannot thrive in them because mosquitoes are numerous there. Avian pox, a virus, can be spread by biting insects like mosquitoes but also by contact and even as aerosols (e.g., in aviaries). As with avian malaria, the frequency of pox lesions declines to almost nothing at high elevations, suggesting that mosquitoes are the key transmitter.

A concern is that, as global warming proceeds, high elevation areas will become warmer and mosquitoes will then invade them. This change will, in turn, restrict the birds to ever smaller areas of malaria-free forest and may ultimately lead to their extinction. If the average temperature in Hawaii were to increase by ca. 2°C in this century, as climate models predict, then 60–96% of the high-elevation refuges from disease would disappear. On the lower islands, like Kauai, probably all the high-elevation refuges would vanish, along with all the native bird species. However, one finding inspires very cautious hope. One native Hawaiian bird species—the Hawaii amakihi—has evolved substantial resistance to avian malaria. So far, though, this is the only Hawaiian species that has managed such a feat, and the small population sizes of the remaining native bird species make it doubtful that they will be able to evolve resistance before they disappear entirely.

The plight of the Hawaiian native birds is just one of many cases in which an introduced species carries a pathogen to which it is relatively resistant but that devastates susceptible native species. As noted already, the North American gray squirrel outcompetes the native red squirrel in Great Britain and is also favored by the widespread replacement of conifer woodlands by those dominated by deciduous trees. However, this effect is compounded by the fact that the gray squirrel harbors a disease known as squirrel pox. Although the invader is resistant to the disease, the red squirrel is highly susceptible; all or almost all individuals infected by the virus die. Similarly, crayfish plague, a water mold discussed in Chapter 2.3, was introduced to Europe in the 19th century either in ballast water or in transported North American crayfish. To compensate for the decline of the native noble crayfish, partially caused by crayfish plague, North American signal crayfish were introduced. It was later discovered that they carry the disease but are resistant. The signal crayfish has largely replaced the noble crayfish in parts of Europe, and the disease is a major factor. Subsequently, the North American red swamp crayfish was introduced to Europe, and it also carries crayfish plague. A similar scenario played out in Atlantic coastal regions of the United States, where decline in the native eastern oyster from habitat degradation, overfishing, and in some areas the introduced protozoan oyster diseases MSX and dermo spurred introduction of the Pacific oyster from Japan in 1950. However, the Pacific oyster also carries the protozoans, and movement of the Pacific oyster to various sites even helped spread the diseases. Thus, the introduction exacerbated the decline of the eastern oyster.

A current worldwide threat to frogs is the introduced fungal disease chytridiomycosis, first detected in the early 1980s by researchers on various frog and salamander species. These scientists began noticing precipitous declines in species they were working on, sometimes to the point of disappearance in just one or two years. One crashing population was that of the famous golden frog, the national symbol of Panama, studied intensively by herpetologists. Similarly, in the mid-1990s, midwife toads disappeared in just three years from a mountain region in which they had formerly been abundant. Many stories like these, of species disappearing before the very eyes of researchers who had long studied them, led to publicity about a global amphibian decline, culminating in an international meeting in 1989. It now appears that the pathogenic fungus is the key culprit in the decline of amphibians, though other factors sometimes play a role and often act synergistically. Today, chytridiomycosis has been diagnosed in 400 of the 6,300 amphibian species worldwide, including most of those that are rapidly declining or have gone extinct recently.

The fungus is thought to be native to sub-Saharan Africa but is now found on all continents but Antarctica. One of the earliest fungus-infected museum specimens is an individual of the well-known African clawed frog species. For years this frog was used as a pregnancy assay in humans after it was discovered that injection of urine from pregnant women induced ovulation in this frog. It was also widely used in laboratories worldwide to study mating reflexes because it could be kept in captivity easily. It was therefore shipped in great numbers to other parts of the world, primarily for breeding to create populations used in pregnancy tests. Even after the development of nonbiological pregnancy tests in the 1970s, these frogs continued to be shipped overseas from Africa as models for studies in immunity, embryology, and molecular biology. The African clawed frog itself resists the disease, having coevolved with it, so it was not obvious that these frogs carried a pathogen. In the importing country, escaped frogs, the water they lived in, or both must have come in contact with local amphibian species and transmitted the fungus. Once chytridiomycosis was established in other species, they could have spread it, because people ship other frogs around the world. One likely later vector is the American bullfrog, which is shipped for the food trade and to establish populations for harvest.

Whirling disease (Chapter 13) in trout is caused by a European protozoan-like parasite that penetrates the head and spine of juvenile fish, where it multiplies and exerts pressure on the organ of equilibrium. The fish then swim erratically and have difficulty feeding and avoiding predators. The parasite has a complex life cycle. Spores reach the substrate when an infected fish dies or is eaten by a predator (in which case the spores are expelled in feces). In the substrate they can survive freezing and drying for up to 30 years. Spores must then be eaten by a sludge worm, in whose intestine the spore develops into a form that enters the water, where it contacts young trout; trout can also eat infected worms. Rainbow trout are particularly susceptible to whirling disease, which reached North America in 1955 by a tortuous route and has since spread widely in the United States. North American rainbow trout were transplanted in Europe in the 19th century, and whirling disease quickly infected all European populations. Rainbow trout probably got the pathogen from brown trout, a European native that harbors the parasite but resists the disease. Eventually, frozen rainbow trout from Europe were exported to the United States and found their way to grocery stores in Pennsylvania, where a stream flowing through a residential area probably carried the parasite to a nearby fish hatchery. Trout transfers from this hatchery spread the parasite to many other states. Whirling disease has been an economic disaster in parts of the American West; in many streams in Montana and Colorado, whirling disease afflicts over 95% of the rainbow trout, devastating the sport fishery.

Bats in North America are currently gravely threatened by a pathogen that was almost certainly introduced on the shoes or clothing of someone visiting a cave who had recently been in European caves. In 2006, a visitor to a popular cave near Albany, New York, noticed several hibernating bats with white noses and others dead on the cave floor. The next year, similar reports came from other caves in the vicinity, as did observations of bats awakening from the torpor that typically occurs during hibernation. These bats were found to be infected by a fungus native to the Old World that causes the disease bat white-nose syndrome. Afflicted hibernating bats often awaken and fly in the winter, using up energy reserves without capturing insects. In some caves, over 90% of bats die. The fungus has already killed over a million bats of six species, including the endangered Indiana bat. Several other bat species, including endangered ones, appear to be at risk as the fungus spreads. Mutual grooming among bats and the high bat densities in caves aid the spread of the disease. In Europe, where the fungus is native and most bats are resistant to the malady, it is not nearly so devastating. The fungus cannot survive in warm temperatures, which may ultimately limit its range in North America, although it has already spread north to Quebec and Ontario and south and west to North Carolina, Oklahoma, and Missouri. Now that the fungus is established in North America, government agencies and bat conservation organizations are urging caves to close or to limit visits (resisted by some owners of commercially visited caves) as well as to disinfect clothing and equipment that have been in caves with infected bats.

Introduced parasites of plants can be disastrous for agriculture. For instance, parasitic witchweed from Africa arrived in the southeastern United States after World War II and inflicts great losses on grass crops (including corn); it has been the focus of a long eradication campaign (Chapter 9.3). The bacterium from Asia that causes citrus canker repeatedly infests citrus groves in other parts of the world, causing devastating losses. A series of invasions of Florida in the late 20th century led to the deliberate cutting and burning of over 10 million citrus trees and cost the industry hundreds of millions of dollars. These actions by the industry and state agencies also engendered enormous ill will from thousands of residents whose healthy ornamental citrus plants were destroyed in a failed effort to eradicate the disease. Introduction of a modified form of a pathogen that has previously invaded can also greatly affect agriculture. For instance, the water mold that causes potato blight, which probably originated in the Andes, has been in the United States since the early 19th century and had long been controlled by fungicides. However, a fungicide-resistant strain entered the United States from Mexico in the late 20th century, spread throughout the East and Midwest, and caused substantial crop losses as well as costs of tightened harvesting, processing, and equipment-cleaning procedures.

The complexity of some interactions between introduced parasites or pathogens, vectors, and new native hosts can be impressive. Chinese grass carp were deliberately introduced to Arkansas in 1968 to control introduced aquatic plants; no one realized they were infected with an Asian tapeworm. The grass carp subsequently invaded the Mississippi River, where the tapeworm infected native fishes, including the red shiner, a popular baitfish. Fishermen or bait dealers then introduced infected red shiners to the Colorado River, and by 1984 they had reached a Utah tributary of the Colorado, the Virgin River. In the Virgin River, the tapeworm ravaged populations of woundfin, a native minnow already threatened by dams and hydrological projects.

Introduced pathogens and parasites are not always accidental; they have also been used intentionally in biological control of introduced hosts, often to great advantage. For instance, an introduced South American parasitic wasp has partially controlled the highly damaging South American cassava mealy bug in Africa. The risk, however, is that biological control agents and their targets are living organisms, and they can evolve and thereby affect the interaction between them in unforeseen ways. For instance, the myxoma virus was introduced from the Americas to control rabbits in continental Europe (where the European rabbit is native) and Great Britain and Australia (where it is introduced). Initially the project was a great success, with rabbit mortality over 90%, but eventually the virus evolved to be more benign. Now, in Great Britain and Australia the rabbits also have evolved increased resistance to the virus, so that, in each successive epizootic, mortality is lower.

3.6 How does hybridization with invaders affect native species?

Many introduced species have hybridized with related native species, sometimes producing a genetic change so great as to constitute a sort of "genetic extinction" of the latter. In some cases the hybrids can mate with one another but can also backcross—that is, mate with individuals of the native population. If the size of the introduced population is larger than that of the native population, the resulting "hybrid swarm" of offspring is a mixture of genotypes that tends increasingly to be dominated by individuals mostly descended from the introduced species. For this process to proceed, the introduced and native species must be closely related genetically so their hybrid offspring will be fertile and able to backcross into the native population. In a region that has been very isolated from others for a long time, the native species are so evolutionarily distinct that they will not be able to mate with invaders. An example is the marsupials of Australia currently faced with invasion by placental mammals. By contrast, many European and North American species are quite closely related to one another, and within single continents (such as the eastern and western sides of the Rocky Mountains in North America) the relationship may be even closer, so that hybridization and backcrossing are more likely.

A good illustration of this problem is the threat faced by the endangered European population of the white-headed duck. Originally common in the Mediterranean region, the white-headed duck was reduced by habitat destruction and hunting to just a few dozen individuals in Spain. Protection and captive breeding have increased this Spanish population to perhaps 2,000 birds. However, this remnant population is now threatened with genetic extinction by the North American ruddy duck, which was first kept in captivity as a curiosity in England in 1949 but quickly escaped. The ruddy duck population thrived and spread, reaching first France and then Spain, where it hybridizes with the white-headed duck population, with backcrossing into the latter. As both ruddy and hybrid males are socially dominant over white-headed males during courtship, the genetic threat is great. An effort has been mounted to eliminate the ruddy duck and hybrids in Spain, but it is hindered by the fact that many hybrids are visibly almost indistinguishable from white-headed ducks.

Mallards, native to North America and Eurasia, have been widely introduced for hunting and frequently hybridize with related native ducks, in several instances exacerbating conservation problems. For example, competition with introduced mallards has led to a population decline of the endangered New Zealand gray duck, and the remaining individuals are increasingly hybridizing with mallards, with backcrossing into the gray duck population. Hybridization with introduced mallards similarly contributes to the decline of the endangered, endemic Hawaiian duck and interferes with projects to save the native species by reintroducing it in several sites. Introduced mallards also hybridize with endemic native ducks in Florida, Madagascar, and Africa.

Hybridization is believed to have played a role in 38% of all native fish extinctions in North America during the 20th century. Fish introduced as game fish, baitfish, and biological control agents have all played roles in these extinctions. In one well-known case in Texas, native Amistad gambusia hybridized with mosquito fish introduced to control mosquitoes, leading to the genetic extinction of the native. Several native trout in the West, including the Apache trout and the gila trout, both listed under the U.S. Endangered Species Act, are threatened partly because they hybridize with rainbow trout, which, although native to the West, have been introduced to new regions for fishing.

The same phenomenon can be found in all groups of species. Among insects, the native southern blue butterfly of New Zealand is threatened with extinction by increasing hybridization with its Australian relative, the common grass blue. In South Florida, the endemic pineland lantana is gradually being absorbed into a hybrid swarm by hybridization with the popular ornamental plant West Indian lantana. Similarly, the rusty crayfish, native to the Ohio River Valley, has been introduced elsewhere in eastern North America for fishing and is now extensively hybridizing with the native northern clearwater crayfish, threatening this species with genetic extinction.

Gene flow between an introduced species and a native species can threaten the latter even when the two species do not hybridize! In the early 20th century, the white sucker, a fish native to the east side of the North American Continental Divide, was introduced to the Colorado River (west of the Continental Divide), where the flannelmouth sucker and the bluehead sucker are native and co-occur without hybridizing. However, both native species now hybridize with the introduced white sucker, leading to indirect gene flow between the two native species and producing a hybrid swarm with genes from all three species.

Sika deer from Japan were introduced to Great Britain between 1860 and 1930 and established wild herds. Although they are much smaller than the native red deer, hybrids between the two species are fertile and can backcross. Recent research shows that, in parts of Scotland, 40% of what had been believed to be red deer are in fact to varying degrees hybrids with sika deer. Even though hybridization between the two species occurs only once in every 500 to 1,000 matings, this degree of gene flow has sufficed to cause the red deer to evolve to become smaller and the sika deer to evolve to become larger. Similarly, feral housecats now interbreed with wildcats to the extent that, in northern and western Scotland, believed to have the purest wildcats in Europe, 80% of individuals have traits of domestic cats. A similar problem is occurring in southern Africa, where feral housecats are hybridizing with African wildcats.

Throughout the 1960s, hybrids were generally recognized by phenotypic features—that is, by their appearance, which was often midway between those of their parental species. However, recent advances in molecular techniques have led to an explosion of research on the genetic constitution of populations of many species. This research has shown that a great many cases of hybridization had been occurring between introduced and native species with little or no visible, phenotypic trace. The hybrid white-headed ducks in Spain are just one example.

Hybridization between introduced and native species has even produced new species. Perhaps the most famous case is the one mentioned in Chapter 1.3, the cross between smooth cordgrass, introduced to the United Kingdom in the mid-19th century from coastal eastern North America, and native small cordgrass. The resulting hybrids were sterile because of chromosomal incompatibilities. However, around 1890 one of these hybrids underwent a chromosomal mutation—a doubling of chromosome number—and instantly became a fertile, sexually reproducing new species, known as common cordgrass. Similarly, Oxford ragwort, itself a hybrid of two Italian species, escaped from the Oxford Botanical Garden around 1690. During the Industrial Revolution it spread through much of Great Britain along railroad lines, producing sterile hybrids with several related native British species. However, eventually a chromosomal mutation (again, a doubling of chromosome number) in a hybrid between Oxford ragwort and native groundsel generated a fertile new polyploid species, Welsh groundsel. Aside from plant and animal hybrids, hybridization has also played a major role in generating new and more virulent species of plant pathogens, as has already been discussed—for instance, among the previously mentioned water molds and bacteria.

Hybridization with invaders can even threaten native species when no gene flow occurs. European mink populations have plummeted because of habitat destruction and pollution in much of their range, and American mink have been brought to Europe as a furbearer. American mink have escaped in several regions and established populations. American mink males become active well before European mink males and mate with female European mink. The females eventually abort all hybrid embryos, so gene flow does not occur; however, these females are removed from the European mink breeding population for a season, which is a major handicap for a small, declining population.

4

IMPACTS OF INVASIONS— COMPLICATIONS AND IMPACTS ON HUMANS

4.1 What is an indirect effect?

Whereas impacts like predation and herbivory are direct effects of an introduced species A affecting a native species B, impacts of invasive introduced species on native species are often mediated in a number of ways by other species, which may themselves be native or introduced.

In a phenomenon known as *apparent competition*, a species causes decline of another species not by actually competing with it for a resource or interfering with its access to a resource but rather by constituting a food source for a predator that then preys on a second species, reducing the population of the latter. An example is how the introduced snowshoe hare affected the native mountain hare in Newfoundland. The latter inhabited Newfoundland's boreal conifer forests until the snowshoe hare was introduced for hunting in 1870. The snowshoe hare is the common North American boreal forest species, and since its arrival in Newfoundland the mountain hare is found only in tundra. This appears superficially to be a simple case of competition, but it is not. Rather, the population of lynx, native to Newfoundland, spiked as the snowshoe hare population increased because the lynx had more food. This in turn put great pressure on the mountain hare, also a lynx prey species, and it found refuge in tundra, where boulders and other features can be used as cover from the lynx. A similar example has been studied in California vineyards, in which the arrival of a new plant pest, the variegated leafhopper, led to decline of the native grape leafhopper. This decline was caused not by competition but by the increase in populations of a native parasitic wasp, the fairyfly, because of the new host species, the variegated leafhopper, but this increase negatively affected the grape leafhopper population. In another case, in the Gulf of Maine, the European green crab enables the European common periwinkle snail to dominate a native relative, the rough periwinkle, because the latter species is far more susceptible to crab predation than the former one. An introduced herbivore can similarly generate indirect competition between two species it feeds on. Ecologists Jennifer Lau and Sharon Strauss have shown in California that burclover, introduced from the Mediterranean basin, serves as a food source for the introduced Egyptian alfalfa weevil, increasing the population of the latter and causing it to feed more intensively on the native Chilean bird's foot trefoil.

There are many other indirect effects of introduced species in addition to apparent competition. One of the most common indirect effects is the "trophic cascade," in which introduced species A eats native species B, thereby reducing the population size of B and lessening its feeding on population C. The cascade can even extend to a fourth species if the resulting increase of population C leads to increased feeding on population D, which then declines. The alewife invasion of the Great Lakes (Chapter 3.2) became part of a trophic cascade once chinook and coho salmon were introduced from the Pacific Northwest. Recall that the alewife, by eating and reducing populations of zooplankton, contributed to the decline and extinction of native fish species that also fed on the zooplankton. However, predation by salmon has now greatly reduced the alewife population and has thereby led to an increase in zooplankton populations. Similarly, a good example of a trophic cascade is the precipitous decline of native fisheries in the Black Sea with the introduction of the sea walnut and the dramatic fishery recovery brought about by the subsequent introduction of another comb jellyfish that preys on the sea walnut. Trophic cascades are also the basis of many biological control projects, in which a predator or parasite is introduced to control an introduced herbivore. An example is the use of the South American lady beetle to control the orthezia scale, which was destroying gumwood trees on St. Helena Island (Chapter 3.3). Other examples will be discussed in Chapter 10.3.

An interesting trophic cascade occurred on subantarctic Macquarie Island, a World Heritage site discovered in 1810. This case demonstrates how indirect effects such as trophic cascades can lead to unintended harmful consequences. Sailors introduced cats to control rats and mice that they had inadvertently introduced and that threatened their food stores. In 1878, sealing gangs introduced rabbits as a food source. The rabbits thrived and also proved to be a common prey item for the cats. The rabbits caused enormous damage to the vegetation, so the rabbit myxoma virus was introduced in 1968. Rabbit numbers quickly plummeted, and vegetation began to recover. However, the cats then turned to native ground-nesting birds as alternative prey, spurring a cat eradication program initiated in 1985. All cats were eliminated by 2000, but then the rabbit population exploded and again devastated the vegetation. Thus, through a trophic cascade, introduction and then elimination of cats led to a decline in vegetation.

Unforeseen impacts of intentional introductions can arise from a variety of indirect effects, some of which are so idiosyncratic that they cannot be classified as trophic cascades or apparent competition. These chain reactions can be surprising. For example, in Great Britain, caterpillars of the native large blue butterfly develop in underground nests of a native red ant, where they parasitize the ant by feeding on its larvae before pupating. The ant cannot nest in overgrown areas, and for many centuries there was no dearth of cleared areas because of grazing and cultivation. However, rabbits became the chief grazers because of a decline in livestock grazing and changing land use patterns during the 19th century. In 1953, South American myxoma virus, which had devastated rabbits in Australia, was illegally imported into Great Britain and spread both by itself and through illegal movements by landowners wishing to rid their estates of rabbits (which are themselves introduced, not native, in Great Britain). The virus devastated the rabbit population, the ant population crashed, and the butterfly population subsequently dwindled to extirpation in Great Britain. More recently, this butterfly has been successfully reintroduced from Sweden and maintained by grazing plus planting of thyme, its host plant.

Another unexpected chain reaction began in 1916 when kokanee salmon were introduced to Flathead Lake, Montana. They largely replaced the native cutthroat trout and became a popular sport fish. The kokanee dispersed from the lake, and their populations became so large that when they spawned they attracted many predators, including bald eagles and grizzly bears. In the 1970s, opossum shrimp were introduced to the Flathead catchment as food for kokanee, and they reached Flathead Lake by 1982, causing an immediate crash in the populations of their prey, tiny aquatic arthropods. The population of kokanee, competing with the opossum shrimp for these prey species, crashed, causing a major decline in numbers of bald eagles, grizzlies, and other predators.

Many introduced species have indirectly devastated native species by bringing and hosting novel pathogens to which they are resistant but to which one or more native species are susceptible. The Asian birds that brought avian malaria and avian pox to Hawaii, the signal crayfish and red swamp crayfish that carried crayfish plague to Europe, and the gray squirrel that introduced squirrel pox to Great Britain are all examples. Other introduced species indirectly affect native species by vectoring virulent pathogens. For instance, the southern house mosquito from North America transmits avian pox and avian malaria from introduced birds to native Hawaiian birds.

4.2 What is invasional meltdown?

Two or more introduced species can interact with one another to exacerbate the impact on native species, communities, and ecosystems. This phenomenon is termed "invasional melt- down." The devastating outcome of the introduction of Old World thistles to the Argentinian pampas, exacerbated by introduced livestock (Chapter 2.4), is an example. Often an introduced species remains quite innocuous in its new home until another introduced species invades, and then the prior species becomes dramatically more problematic. For instance, in south Florida, ornamental fig trees were common for at least a century, restricted to sites where they were planted by humans because they could not reproduce without fig wasps—their only pollinators. In general, each fig species can be pollinated by only one fig wasp species, and each fig wasp species pollinates just one fig species. Thus, ornamen- tal fig species in Florida were effectively sterile and therefore noninvasive without their fig wasps from their native ranges. However, the fig wasp of the Asian fig known as Chinese banyan recently managed by unknown means (surely hitch- hiking on living plants or other cargo) to reach Florida, and Chinese banyan, now that it is fertile and reproducing, has become invasive, including in natural areas.

In other invasional meltdowns, an introduced species that is already a pest becomes a worse one by virtue of a subse- quent introduction. The impact of an invasive plant species, for instance, is often exacerbated, or at least accelerated, by introduced animals that disperse its seeds. For example, seeds of the nitrogen-fixing firebush in Hawaii, discussed already, are mainly dispersed by the introduced Japanese white-eye, introduced feral pigs, and introduced rats. A further melt-down occurs because the density of exotic earthworms under the firebush is at least twice that under the native vegetation it is replacing. Exactly who is facilitating whom in this case is unclear, and the plant and the worms could be facilitating one another. Whichever is the case, the increased earthworm den-sity with the spread of firebush is helping to bury the nitrogen in the firebush leaves in the soil and is thus increasing rates of nitrogen cycling. This activity is leading to invasion by exotic plants that had been unable to thrive in the nitrogen-poor soil.

Introduced earthworms play roles in other meltdowns. For example, in New Jersey, in an interaction between European earthworms and two invasive plants, Japanese barberry and Japanese stiltgrass, there are much higher densities of the worms in the soil under these invasive plants than under the natives. Nitrogen cycling is therefore greatly sped up, acceler-ating the important ecosystem process by which nitrogen is transformed into a form plants can use. This complex of these two invasive plant species plus the invasive worms produces conditions, especially elevated nitrate, that promote invasion by other invasive plants.

Often an introduced species modifies the habitat to favor a second invader. For instance, through its filtering activities and modification of the substrate, the zebra mussel enhances populations of invasive Eurasian watermilfoil, an aquatic plant introduced to North America in 1942. The direct impacts of Eurasian watermilfoil, such as overgrowth of native plants and interference with swimming and recreational boating, make it one of the most troublesome aquatic invaders of North America. However, Eurasian watermilfoil also facilitates growth of zebra mussel populations by providing additional settling substrates, and it can help disperse zebra mussels between water bodies when fragments of it break off (or are caught by boat propellers) and float or are carried to new sites.

The mussel, in turn, aids the watermilfoil by filtering the water and increasing its clarity. Thus, a mutualism between two damaging invaders worsens impacts of both. In another aquatic invasional meltdown, an indirect effect between two Atlantic species introduced to San Francisco Bay exacerbates the impact of one of them. The amethyst gem clam had been present for over 50 years but was not always very common, as it was outcompeted by native clams. The European green crab changed this situation drastically by preying on the natives and thus releasing the gem clam from this constraint. Simply by providing food for a previous invader, a newly introduced species can exacerbate the invasion. In Spain, ecologist Yolanda Melero has shown how the invasive American mink population has grown in the wake of the introduction of the red swamp crayfish, which dominates the mink diet when it is present. There is no evidence that the mink substantially reduce the crayfish population.

Ants and scale insects are often associated in cases of invasional meltdown, because scale insects secrete a substance called honeydew that ants like to eat. Many ant species are adapted to tend scale insects, moving them around and protecting them from predators. Even the California red scale (in fact an invader from Asia), which does not produce honeydew, is protected and transported by the Argentine ant, making both species more problematic. On the island of St. Helena, introduced ants are interfering with the otherwise successful control of a scale insect by an introduced beetle (Chapter 3.3) by attacking the beetle in some locations, thereby protecting the scale insect.

A famous meltdown involving invasive ants and scales occurred on Christmas Island in the Indian Ocean. Here the legendary native red land crab was being destroyed by an outbreak of the introduced yellow crazy ant, believed to have killed over 20 million red land crabs before a rescue operation launched in 2002 (Chapter 10.2). The feature of this case relevant to invasional meltdown is that yellow crazy ants had already been on Christmas Island for around a century without apparent major impact. They were first detected in great numbers in 1989 and then quickly formed enormous colonies known as supercolonies, consisting of millions of ants, that began to eliminate the dominant land animal, the red crab. The ants are greatly facilitated by, and tend, two honeydew-producing scale insects, at least one of which, the green coffee scale, is introduced and appears to have arrived slightly before the ant population exploded. The honeydew also fosters the growth of a sooty mold that causes tree canopy dieback. Furthermore, the crabs are the main herbivore and also decomposer of plant material. When the ants killed the crabs, seedlings that had been killed by the crabs began to survive at rates elevated many-fold, and leaf litter built up. In sum, the invasive ants and scales generated a complex meltdown involving many species of plants and animals, including soil organisms.

Invasional meltdown can also exacerbate impacts of introduced agricultural pests. Soy, introduced to North America from Asia, is plagued by two such meltdowns. Kudzu, the legendary invasive Asian vine that has come to symbolize the American South, is an alternate host for soybean rust. The rust is an Asian pathogenic fungus now found in many parts of the world that costs soy farmers in the United States hundreds of millions of dollars in annual losses, as rust-induced defoliation reduces soybean yield and can even kill plants. In addition, since its discovery in Wisconsin in 2000, the Asian soybean aphid has become one of the most costly agricultural pests in North America. Its impact on soy is magnified by the fact that the aphid can overwinter on common buckthorn, its alternate host, itself a common, damaging, invasive Old World shrub or small tree.

4.3 Why do many impacts occur only after a time lag?

Introduced species sometimes remain innocuous and restricted in range or habitat for decades or longer and then suddenly expand to become serious pests. Alarmingly, this means that many places contain introduced species not currently seen as problematic that are destined to become pests in the future.

For instance, several cases of invasional meltdown, such as those featuring the figs, the gem clams, and the yellow crazy ant, occurred decades after the main players invaded. In these cases, the long time lag before the big impact is not mysterious; it was simply a matter of a second introduced species arriving, as when the Chinese banyan needed its pollinating fig wasp to reproduce. These are just some examples of situations in which an introduced species remains innocuous and relatively dormant in the environment for a long time and then undergoes a rapid population explosion to become a raging pest.

In many cases a long time lag ends and a species begins to spread for reasons that have nothing to do with the arrival of another introduced species. A good example is the mutation that caused sterile hybrids between small cordgrass and smooth cordgrass in England to become a fertile, new, and invasive species, now known as common cordgrass. Sterile hybrids had been observed occasionally for many decades before the mutation occurred in one of them. Even more remarkably, sterile hybrids of the same two cordgrass species have recently been discovered in France, and their origin is completely independent of the introduction of smooth cordgrass to Great Britain. However, these French hybrid clones have not yet undergone a mutation to become a sexually reproducing new species.

New mutations are often invoked to explain the termination of such time lags, but such mutations have rarely been documented, as was the case with the chromosomal mutation that produced common cordgrass. Nevertheless, there is strong evidence that mutations can produce an invasive genotype. For instance, it has been shown that the aquarium strain of "killer alga" is cold-tolerant and is thus able to survive the winters of the northwest Mediterranean, while populations from their native Pacific range are not (Figure 4.1). This difference suggests that a mutation may have occurred during about 15 years that the alga was cultured in aquaria before it was released to the wild from the Oceanographic Museum of Monaco. However, there was no lag in the spread of the killer alga once it reached the Mediterranean.

By contrast, sometimes a subtle change in the biotic or abiotic environment causes a sudden population explosion of a previously harmless introduced species. Brazilian pepper, for example, was present but not invasive for half a century in south Florida before rapidly expanding to become the most widespread invasive plant in the state beginning in the 1940s (Chapter 3.1). This expansion has been plausibly linked to several human modifications of the environment, including a progressive lowering of the water table because of withdrawal for agriculture and other human uses as well as high soil nutrient concentrations from fertilizer and breakage and plowing of rock for agriculture. A case surely caused by an environmental change is the invasion by a wood-boring isopod, the

Figure 4.1 Pacific killer alga overgrowing seagrass meadows in Mediterranean. (Photograph courtesy Alexandre Meinesz.)

gribble, in Los Angeles Harbor. Introduced in the 19th century, probably on wooden boat hulls, it was either absent or scarce in most of the harbor because of heavy pollution from industrial, domestic, and storm water sources. A successful pollution abatement program in the late 1960s led to a population explosion of such magnitude that several local wharves collapsed.

However, many lags, and their subsequent termination, remain mysterious. Possibly the most dramatic example is that of a Japanese fungus, *Entomophaga maimaiga*, released in the United States in 1910–1911 in an attempt to control the gypsy moth. Unrecorded for the next 79 years, the fungus surfaced again in 1989 and now attacks several gypsy moth populations in the northeastern United States. This lag is so mysterious that some researchers suspect that the original population died out and the later one is a new introduction. In another curious case, giant reed was introduced to southern California from the Mediterranean region in the early 19th century as a roofing material and for erosion control. It remained locally present but not invasive and problematic until the mid-20th century, when it spread dramatically and damaged wetlands, fostered fires, and spurred several ecosystem changes. The reason for this switch is unknown. However, that these mysterious times lags did occur is undoubted; it is the causes that remain undetermined.

In several instances, the use of modern molecular genetic techniques has shown that the genetic diversity of a species is greater in an invaded region than in the native range. This oddity reflects introduction of populations from different source areas, and sometimes this genetic enrichment is what terminates an invasion lag. For instance, the Caribbean brown anole lizard was introduced to extreme south Florida beginning in the 19th century. However, in the 1940s its range began to expand, and the rate accelerated in the 1970s so that it now occupies most of Florida. The Florida population has far greater genetic diversity than any native population does, a result of several independent introductions likely from different parts of Cuba. It is possible that mating in Florida between individuals descended from different source populations produced a more invasive genotype or that one of the late introductions happened to possess a genotype particularly well suited to a wider range of environmental conditions than its predecessors.

4.4 Do invaders ever just go away on their own?

Sometimes explosive, damaging invasions quickly collapse for unknown reasons, and the introduced species persists as a less prominent or even innocuous resident in its new home. The classic example is the trajectory in England of Canadian waterweed. First seen in a pond near the Scottish border, it quickly spread to rivers, canals, ditches, and ponds throughout much of Great Britain. At the height of the invasion in the 1860s, it clogged the Cam River to the extent that it interfered with rowing, and extra horses were required to tow barges. At least one bather drowned after becoming entangled in it. Parts of the Thames were impassable. Then, without human intervention, Canadian waterweed suddenly declined to a moderate or even lesser status throughout its British range by the early 20th century. More remarkably, similar collapses of the same invader occurred in Sweden and Germany. The reason for these collapses remains unknown. Another mysterious, apparently spontaneous collapse is that of the invasive giant African snail on many Pacific islands. The tragedy in this case is that the rosy wolf snail was introduced to control the giant African snail, but the latter proved to be too large, while the predator instead attacked many native snails, driving some species to extinction (Chapter 3.3). In inexplicable cases such as the crashes of Canadian waterweed and giant African snail populations, it is tempting to suspect that a pathogen either arrived from elsewhere or became epidemic once the host population achieved a certain density, but in neither instance has such a pathogen been found.

Another well-known collapse was of the crested myna, introduced from Asia to the city of Vancouver, British Columbia, in 1897. After a short lag, this bird spread rapidly throughout southwestern British Columbia and as far south as Oregon by the early 1930s, eventually numbering some 20,000, including over 6,000 birds in Vancouver alone. The population then dwindled, and the last two individuals died in 2003. Many possible causes have been suggested, including competition with starlings that invaded in the 1950s, but there is little evidence to support any of these hypotheses. Similarly, budgerigars were commonly brought to the United States in the first half of the 20th century as cage birds; some escaped, and others were released. A substantial breeding population established in Florida on a large swath of the central Gulf coast, aided by nest boxes. Peak numbers in the 1970s were around 20,000. However, by the mid-1990s they had disappeared entirely from all but about 20 miles of coastline, and by 2000 only about 200 individuals remained. Competition for nest sites with introduced house sparrows and starlings has been suggested as at least one factor in the decline, but no strong evidence supports this claim.

Substantial detective work has elucidated other population crashes that had initially seemed mysterious. For instance, the European browntail moth invaded Massachusetts at the end of the 19th century and quickly spread through New England and into Canada. Its huge populations and the fact that it attacked many tree species led to great fears that it would outdo the invasive gypsy moth. However, populations crashed in 1913–1914, and it disappeared from Canada by 1927. By the 1970s it was found only on a few islands in Casco Bay, Maine, and some coastal sites near the tip of Cape Cod. In this instance, research by entomologist Joseph Elkinton and colleagues showed that this collapse, long viewed as puzzling, was probably not autonomous but rather a fortuitous unintended consequence of the introduction of a European parasitic tachinid fly species to attack the gypsy moth.

A current notable example of spontaneous regression of an invader is the decline of the "killer alga" from the south Pacific, which reached the Mediterranean in the 1980s in water that was not properly disposed of from public aquaria at the Oceanographic Museum of Monaco. The alga quickly spread to the coasts of France, Spain, Italy, and Croatia, covering thousands of acres of nearshore substrate and overrunning the seagrass meadows that support many marine fishes and invertebrates. Commercial and sport fishing were curtailed, and many other ecological impacts have been detected. Over the last few years, however, the alga has greatly declined throughout the Mediterranean, to the point that it is now difficult to find in some areas it had come to dominate. The reason for this collapse is unknown. One possibility is related to the fact that the entire invasion appears to consist of a single male clone, which spreads by vegetative growth and by pieces that break off (e.g., when they are severed by boat propellers) and float elsewhere. No female gametes have been detected in the Mediterranean. Absence of sexual reproduction in any species is believed to inhibit evolutionary responses to changing conditions, such as presence of new diseases or modified physical environment. However, scientists have not documented any obvious changes that could have spurred the decline of this alga. It is also possible that this regression is simply part of a long-term cycle and that the killer alga will spread once again in the future. In any event, the environmental benefit deriving from its regression is dampened by the inexorable spread of a closely related species, sea grape, also native to the western Pacific and introduced somewhat later than the killer alga (Chapter 2.6).

The fact that some invasions spontaneously collapse should not be a guide for policy decisions (Chapter 9.2). Not many such collapses are known, certainly far fewer than the number of cases of invasions after a quiescent time lag. Further, even if an invasion collapses, it may leave a long-term, even permanent legacy. For example, an introduced plant can leave a chemical in the soil that persists long after the plant has disappeared and that hinders other plants, a process termed *allelopathy* (Chapter 3). Or an introduced species may die out but leave behind devastated populations of whatever species it was eating. In one such case, a small population of reindeer introduced to St. Matthew Island in the Bering Sea exploded, then overate their food supply, crashed, and eventually died out completely. However, lichens, their main food, are extremely slow-growing, and it will be decades before the traces of this invasion disappear.

4.5 What are the economic impacts of biological invasions?

Ecologist David Pimentel and colleagues have estimated that the United States economy loses about $120 billion annually to invasions in terms of damage and management costs. Losses to agriculture top the list—introduced agricultural weeds cost $24 billion each year in losses and $3 billion in management costs. Introduced rats are estimated to cause $19 billion in damage and losses each year, mostly to agriculture. Introduced insect crop pests add another $14 billion in costs per year, while introduced crop pathogens add $22 billion. Introduced forest pathogens and insects are believed to cost $4 billion annually, and introduced weeds, insect pests, and pathogens of lawns, gardens, and golf courses impose about $5 billion in annual costs. Some invaders generate staggering costs beyond those just cited. For instance, the red imported fire ant, zebra mussel, and Asiatic clam each impose damage and control costs of $1 billion every year. For the zebra mussel and the Asiatic clam, the majority of this cost is incurred by governments, power companies, and industries that must clear water pipes clogged by these mollusks.

Costs of items like lost crops or labor to clear clogged pipes are straightforward to estimate. Other costs are extremely difficult to measure; this is particularly true when the damaged or lost entity has no market value or when such a value must be extrapolated from indirect measures (Chapter 1.3). For example, Pimentel and his colleagues extrapolated from estimates in two states that feral cats kill more than three million birds per year in each state, plus some data on behavior of feral cats, to suggest that each cat kills eight birds per year. But what is the value of a wild bird? Pimentel and colleagues arrived at $30, based on estimates of how much money each birdwatcher spends per bird observed, how much each hunter spends per bird shot, and how much game-rearing facilities spend per bird reared and released. Concatenating these estimates and extrapolating to the entire United States, they concluded that cat damage to bird populations costs about $17 million, and this does not include damage to other animals, such as reptiles, amphibians, and small mammals, or any damage caused by domesticated cats. Even more difficult to estimate, at least in economic terms, is the cost of the many extinctions of native species caused by biological invaders. Pimentel and his colleagues forswore the effort. They also did not delve deeply into public health costs of invasions, which can be enormous, as was shown in Chapter 1.3 with the examples of West Nile virus and chikungunya. Public health impacts will be discussed further in the next section.

Pimentel and colleagues extended this analysis to the United Kingdom, Australia, South Africa, India, and Brazil and estimated that the aggregate cost of invasions for these five nations is about $210 billion annually. Assuming similar losses worldwide, they then estimated that the global cost of invasions is about $1.4 trillion annually, or about 5% of the summed gross national products for all nations, a staggering amount.

Many introduced species are economically beneficial (Chapter 1.4)—for instance, most of the major food crops in the United States are nonnative. Very few crop plants become invasive. Other sorts of introductions are distinctly mixed blessings; some species that become invasive generate economic benefits as well as costs, and net benefit or cost is not always easy to calculate. About half of all ecologically damaging invasive plants in the United States were originally deliberately introduced as horticultural varieties, and some of them—pampas grass, Japanese pachysandra, and various species of privet, for example—are still sold as ornamental plants, generating income for whoever sells them. In a policy and management context, it makes little sense to consider all introduced species or even all invasive introduced species in the aggregate. It is important to note, however, that many introduced species generate huge net costs, and it would have been far better not to have introduced those species. Failing that, it is important to minimize their negative impacts (Chapters 9 and 10).

Highly detailed analyses of the costs of particular invasions are available for several species. The golden apple snail was introduced from Argentina to Taiwan in 1980 in the hope that it would become a high-quality protein food source for people and perhaps even a pricey export commodity as "escargots." Economist Rosamond Naylor studied the economic trajectory of this introduction as the snail quickly spread throughout much of southeast Asia; she used the Philippines as a case study. Unfortunately, the snail invaded the irrigation networks that underpin plantations of rice and began to feed on rice seedlings, injuring the area's main food supply. One initial attraction of this snail was that it feeds voraciously on many plants, grows quickly to a large size, and breeds rapidly; it was nicknamed the "golden miracle snail" when it was introduced. In retrospect, the same features should have provided warning that, if something were to go wrong, it could go very wrong very quickly. Also in retrospect, it is surprising that the entrepreneurs who introduced it did not consider the fact that it was already a well-known problem in irrigated rice plantations in South America.

The golden apple snail was first smuggled into Taiwan, introduced from Taiwan to Japan by entrepreneurs and then to the Philippines with government endorsement. It was subsequently carried to China, South Korea, Malaysia, Thailand, Indonesia, and Vietnam. Although introduced only to ponds, cement tanks, and backyard pits, it inevitably escaped and was probably also released into irrigation ditches and other waterways. The market value of the golden apple snail quickly plummeted because Asians did not like its taste, and various public health regulations precluded its import into Europe and North America. It spread rapidly through rice paddies, infesting 15% of wet-rice area by 1991. Damage can be staggering, up to 90% of the crop destroyed, with a density of one snail per square foot. Over 10% of farmers in one survey had no harvest at all. Total crop loss for 1990 was estimated at about $15 million, and control costs were similar. The total cost of the invasion for 1990 was 25–40% of the cost of rice imports to the Philippines. Remarkably, this cost for one year was equal to what the government could instead have spent on a rigorous inspection and quarantine program for all agricultural introductions.

In addition to the direct cost to the rice crop, there are a number of indirect costs of this snail invasion. It could, for instance, threaten wild rice populations that would be useful sources of germplasm in the future. The snail also has human health impacts. It is an intermediate host of a lungworm that attacks humans as well as rats and also of trematodes that irritate the skin. The molluscicides that have been widely used in control are highly toxic to fish and threaten farm workers as well. These chemicals also have many nontarget impacts (Chapter 9.2) that may change various food webs, though this impact has not been studied.

Ecologist Erika Zavaleta researched the economic impact of another major invasion: Asian salt cedar in the western United

States (Chapter 1.3). Introduced in the 19th century as an ornamental, as a windbreak, and to control erosion, salt cedar spread and replaced native vegetation in 1,250,000 acres in 23 states, at elevations reaching 8,000 feet. Salt cedar uses more water than native plants; Zavaleta calculated the total annual water loss at 49–106 billion cubic feet. Salt cedar growth and the debris it accumulates also impede river flow, and salt cedar traps sediment. All these features exacerbate floods. A series of floods in Arizona in 1977 through 1979 caused $150 million in damages and called attention to the role of salt cedar. Salt cedar also dries up desert springs and oases. Zavaleta tallied costs these impacts imposed on municipal and agricultural water systems, hydropower generation, and flood control, finding the annual total to be $127 to $291 million. In another economic study, ecologist Glenn Edwards and colleagues have researched costs and impacts of invasive camels in Australia. Among major impacts in addition to harm to vegetation discussed earlier are damage to wetlands through fouling and trampling, competition with native animals for food, damage to sites of cultural significance (e.g., waterholes), and damage to people and vehicles in collisions. The number and variety of expensive impacts are mind-boggling; for instance, the sheep industry maintains a 3,486-mile fence to limit predation by dingoes on their flocks, and camels cause at least $40,000 damage to the fence each year just in the state of South Australia. Edwards and his colleagues could not estimate economic costs for damage that was not borne by a market, but costs such as those of management and lost production were estimated at about $11 million annually.

Of course, the very fact that many invaders are economically and ecologically damaging means that economic benefits can accrue to those who combat them. For instance, anyone employed by a government agency charged with inspection of goods to intercept invaders owes his or her livelihood to the fact that harmful invasions occur. Similarly, workers for government agencies, commercial ventures, and nongovernmental organizations who eradicate invasive mammals on islands (Chapter 9.3) enjoy gainful employment by virtue of the existence of harmful invaders. Additionally, many individuals are developing new technologies to manage invaders, as will be seen in Chapters 9, 10, and 12; some patent these technologies, hoping to garner great economic benefits. All are surely earning their livings. These marginal economic gains do not mean that the phenomenon of invasion in general is a good thing for humankind any more than the existence of diseases is a good thing because people in the medical profession derive good incomes from fighting them.

Economic impacts of invaders can be quite bizarre. The Asian multicolored lady beetle, introduced to North America to control aphids, not only outcompetes and preys on native lady beetles (see previous discussion in this chapter and Chapter 10.3) but also has become a serious contaminant of wines in the United States. These beetles aggregate in damaged grape clusters and are crushed in the winemaking process, imparting an unpleasant odor and flavor to wine even when they are present in very small quantities. Now that this lady beetle has been inadvertently spread to Europe and South Africa (Chapter 5.5), there is great concern that European and South African wines will be similarly damaged. Management costs of invasions can also be surprising. For example, following a recent campaign to eradicate the medfly in California, 14,000 claims were filed by car owners. The charge? The campaign was using aerially sprayed insecticide that damaged car paint. The state of California paid $3.7 million to settle these claims.

4.6 What are the public health consequences of biological invasions?

As noted in Chapter 1.3, invasions have long affected public health. Europeans brought smallpox, influenza, and other diseases to the New World and to many islands, and they brought back syphilis from the New World to Europe, beginning with

Naples. Chikungunya, transmitted by the introduced Asian tiger mosquito on La Réunion and in Italy, is another example (Chapter 1.3). Measles, smallpox, mumps, and influenza arriving with Europeans devastated native Americans. The British are even believed to have intentionally distributed smallpox to the Delaware Indians via infected blankets. Similarly, measles took a terrible human toll when introduced to the Fijian islands, and introduced tuberculosis killed most natives of Tierra del Fuego.

Mosquito invasions in particular have inflicted enormous losses on many populations. Yellow fever spread in the New World only after the arrival in 1647 in Barbados of the yellow fever mosquito from Africa in slave ships. In the New World, the Asian tiger mosquito, through competition with the native eastern tree hole mosquito, makes the latter a more competent vector of the LaCrosse encephalitis virus. It does this by causing the eastern tree hole mosquito to develop with a weaker gut that more easily passes the virus so the mosquito can transmit it to humans. In a final example, malaria epidemics in the New World were initiated when the African malaria mosquito reached Brazil from Senegal around 1930.

Perhaps the best known introduced human disease is the plague pandemic of the 14th century that killed about 30 million Europeans. It was precipitated by the arrival of infected fleas from Asia, though the exact route and arrival site are not known with certainty. The most probable trigger was the catapulting of corpses of plague victims into a besieged Genoese colony on the Crimean peninsula. Five hundred years later, the plague pandemic of the late 19th and early 20th century was probably caused by infected Norway rats stowing away on ships and then decamping in various ports accompanied by the Oriental rat flea, which also bites humans. Similarly, around 1830 cholera reached Europe on war ships visiting ports such as London and Hamburg and later on fast clipper ships.

A series of influenza epidemics is probably today's most evident reminder of the public health consequences of arrival of new pathogens. The "Spanish flu" epidemic of 1917–1919 killed half a million Americans and many more millions elsewhere. Then, in 1957, Asian flu killed over a million people, as did the Hong Kong flu of 1968. More recently, the spread of avian influenza beginning in 2003 and swine flu in 2009 have dominated headlines.

5

EVOLUTION OF INTRODUCED AND NATIVE SPECIES

5.1 What is the paradox of invasion genetics?

A paradox of invasions is the fact that many widespread invasions began with very few individuals. For instance, the entire North American population of an introduced European solitary bee species probably originated from one female. Additionally, several parasitic insects used in biological control of other invaders (Chapter 10.3) were introduced in minute numbers and subsequently spread and became very abundant. For example, stocks reared from just eight individuals of an Australian parasitic wasp, released in California for biological control of pest mealy bugs, established and spread to occupy a large area. The classic example of a small founding population is that of the North American muskrat in Europe. Three females and two males from Alaska introduced in 1905 near Prague, Czechoslovakia, founded a population that within a decade numbered in the millions and had spread into Germany and Austria; muskrats now occupy most of Europe. Similarly, four male and five female small Indian mongooses liberated at one site in Jamaica in 1872 spread to occupy the entire island and were carried by humans to initiate populations elsewhere in the West Indies and Hawaii. Finally, the

European starling was introduced in New York City as about 100 individuals in 1890 and 1891, as part of a project to introduce all the birds mentioned by Shakespeare. This species has since spread throughout the continent to become one of the most common birds in North America.

The paradox arises from an established principle of genetics that states that small populations are at grave risk of extinction for two reasons. The first is the threat of inbreeding depression—the many genetic diseases and defects that arise when close relatives mate, as they must in tiny populations. The second is the problem of *genetic drift*, the loss of genetic variation by chance alone, at rates greatly elevated in smaller populations. Because genetic variation is needed for evolution to proceed, the concern is that small populations will be unable to evolve adaptations to a constantly changing environment. Many examples exist of captive populations in zoos and parks that are gravely threatened by inbreeding depression and genetic drift. Yet widespread invasive populations have often begun with very few individuals. They must have experienced what is known as a *genetic bottleneck*, a loss of genetic variation brought about by great reduction of population size. How can they have survived? General resolutions of the paradox will be discussed later, but first it is important to show that, in fact, many invasive populations, even those begun with small numbers, have evolved in a variety of ways and that this evolution is often adaptive so that they are better able to survive, reproduce, and in some cases become more problematic as invaders.

5.2 How do introduced species evolve?

Consider an analogy: a researcher takes several groups of fruit flies and moves them to new growth chambers in a laboratory. She then subjects the chambers to different environments. She may make one chamber colder than the other, another hotter. She can provide different foods for the different chambers or can subject them to different light–dark cycles to simulate day–night cycles at different latitudes. She continues this experiment for many generations and then tests the fruit flies from different chambers to see if and how they have evolved in response to the different regimes she has imposed. In fact, biologists have conducted experiments like this one with fruit flies (and other organisms) countless times, and the degree and speed of evolution can be amazing. Often the evolution appears adaptive. However, especially when small numbers of individuals are used to begin the experiment, there can be different evolutionary changes in the different experimental populations even if the environment is kept the same in all the chambers. Such evolution, caused primarily by genetic drift and inbreeding, is not too likely to be beneficial and may even be maladaptive.

A population of a nonnative species introduced to a new region is faced with the same situation as the experimental fruit flies are. However, the differences between the original and new environments are not controlled by a scientist; they are generally more complex than in the experiment (that is, food, temperature, and other factors all change), and the environmental differences are rarely even measured. But it is small wonder that evidence of many kinds of evolution abounds among introduced species.

Many introduced species have evolved different size or shape in their introduced range. For instance, the small Indian mongoose, introduced to the West Indies, Hawaii, Fiji, Mauritius, Okinawa, and other islands, has evolved in almost all its new homes to be larger than in its native Asia. Further, the males, always larger than the females in this species, have increased in size even more than the females have. A likely explanation is that in Asia this mongoose coexists with several larger carnivores, such as the gray mongoose. They may well be outcompeted by such species for large prey items, and the males, closest in size to the larger competitors, would be most affected, thus most likely to evolve to be larger when such competitors are absent, as is the case on almost all the islands to which the small Indian mongoose has been introduced. This hypothesis is buttressed by an observation from Croatian islands in the Adriatic Sea, the only islands to which the small Indian mongoose has been introduced that already have an established native small carnivore, the larger stone marten. Here the mongoose is not bigger than in Asia.

The stoat is another small carnivore species whose morphology evolved rapidly after introduction to New Zealand from Great Britain in 1884 to control introduced European rabbits. In New Zealand stoats eat, on average, much larger prey than in Great Britain. New Zealand has few small prey species, and stoats there eat rabbits, European hare, Australian brushtail possums, and rats—all introduced species. In Great Britain, various mouse species dominate their diet. This prey shift is correlated with an increase in the stoat's body size and skull length in New Zealand.

Two very widespread introductions, one in North America and the other in Europe, reveal a remarkable degree of morphological evolution in just a century. The house sparrow was successfully introduced from Europe to New York City in 1852 and spread rapidly throughout North America. In its vast native range in Eurasia, this sparrow differs greatly from site to site in size and shape. By the 1960s, it had evolved approximately as much variation in North America as in its native range, with individuals farther north generally being the largest. Equally impressive is the evolution of the muskrat introduced to Europe. In its native range in North America, the muskrat is famous for great variation in size and has about 16 "subspecies" of different size, each in its own geographic region. Just as the house sparrow has in North America, within about a century, the muskrat in Europe has evolved approximately as much morphological variation in its new home as in its native range. Also as with the house sparrow, larger individuals tend to be found farther north.

The introduction and rapid spread of a Eurasian fruit fly in western North America and South America 25 years ago provided a striking example of rapid morphological evolution and a hint that the evolution in this case is adaptive even if is not yet clear what the adaptation is good for. In Eurasia, wing length of this fly tends to be greater in populations that are farther north. After 20 years, researchers detected a similar trend in the introduced North American populations—larger wings in more northerly populations. But a close examination of wings showed that different parts of the wing contributed to this pattern in the Old and New Worlds. Even more remarkably, recently introduced South American populations of this same fly have evolved the same pattern—bigger wings the farther the population is from the Equator—but with yet a different part of the wing generating the pattern. Exactly why it is adaptive for these flies to have bigger wings in cooler climates is not clear, but the fact that they have evolved this pattern independently in three regions strongly suggests that the adaptation is good for something. This fruit fly is typified by chromosomal inversions—mutations in which part of the chromosome has a reverse order of the genes from the original chromosome. There are varying proportions of inversion types in different populations, and these proportions also vary the same way in all three regions. Like the trend toward bigger wings far from the Equator, this pattern provides strong evidence that the different inversion mutants are more or less adapted to particular environments, even though the nature of the adaptation is not yet understood.

The cane toad, introduced from the Americas for biological control of introduced beetles in cane fields, has spread widely in Australia and is legendary for its numbers and its impact on predators, many of which die upon attacking the toxic toad (Chapter 3.3). The process of invasion on an "empty" continent seems to have selected for larger body size, so an already large toad evolved to become still larger. Cane toads in Australia have also evolved relatively longer legs, which leads to greater biomechanical stresses because of greater propulsive force with each hop. Consequently, about 10% of large adult toads have severe spinal arthritis, and the arthritis is associated with not only the large body size but also the relatively long legs and more frequent movement.

Still other introduced species have evolved different "life history" traits in their new homes. The Asian plant velvetleaf arrived in the United States before 1700 and was innocuous for well over a century. In the past 100 years it has become an aggressive invader in cultivated fields in the Midwest. Responding to competition for light, velvetleaf has evolved different life history strategies depending on the species with which it is competing. When growing with soy, it has evolved a response to light quality that enables it to outgrow soy. On the other hand, there is no way velvetleaf can outgrow the much taller corn stalks, so populations of velvetleaf in corn-dominated regions have not evolved this ability. Another problematic species, the European corn borer, was accidentally introduced to North America in 1914 in sorghum imported for use in manufacturing brooms. It quickly became a pest of corn and other crop plants. Since its establishment, the corn borer has evolved genetically distinct biotypes that differ in host plant preference, performance on different host plants, and number of generations per year. Originally, this insect had one or at most two generations per year, but the South now has a biotype that can progress through several generations annually. A similar example among plants is annual blue-grass, native to Europe. Although it is an annual in its native range, in North America it has evolved perennial biotypes in areas that are intensively mowed, like lawns and golf courses.

Many introduced plant-eating insects have evolved both behaviorally and physiologically to use new host plants. For example, the southern cowpea weevil, native to Africa, now attacks many host plants, all in the legume family, and has become a worldwide pest of legume seeds in fields and warehouses. The original host was almost certainly the wild progenitor of black-eyed pea in Africa, but the weevil has evolved biotypes that thrive on many different crop beans. These different biotypes have genetically based egg-laying preferences related to host plant chemistry, and they also survive better on their preferred host plants. Another example of evolutionary adaptations to new host plants is seen in the western corn rootworm, a beetle native to Central America that invaded North America by the early 20th century and Europe in the mid-1990s. One tactic farmers use to deal with its impact has been to rotate corn with soy. However, this tactic was recently thwarted when a genetically determined biotype evolved that prefers to lay its eggs and feed on soy. Biotypes have also evolved that have extended diapause (resting stages) and so can survive the rotation period with soy and attack the next corn crop. These new preferences and abilities involved significant evolution of new behavioral and physiological traits.

Evolution of an introduced plant pest is also seen in the Colorado potato beetle, one of the most famous introduced insects worldwide. This beetle, in fact, is not originally native to Colorado, nor did it originally attack potato. It is a pest by virtue of a complicated voyage and an evolutionary switch in host plants. When the Spanish conquistadores invaded Mexico in the 17th century, this beetle was living on the central Mexican plateau feeding on burweed, a native relative of the potato. The Spaniards introduced cattle to Mexico around 1680 and began driving cattle north to markets in Texas. Cattle picked up spiny burweed seeds and carried them north, where they stuck to American bison. The beetle followed the burweed. The plant moved even farther north in the early 18th century, reaching the Great Plains by 1819. Meanwhile, the potato did not reach North America directly from its native region, the Andes. No one knows who first brought the potato to Europe, but it certainly reached France by the late 16th century. In 1719, potato seed stock was sent from Ireland to New Hampshire, and by 1820, just as the Colorado potato beetle reached the Midwest (on burweed), so did the potato. By 1859, the beetle had evolved a taste for potato, and, after a time lag, the beetle population exploded, reflecting a genetic change. The beetle has now evolved several different biotypes adapted to many potato relatives, including tomato and horse nettle.

One of the most common responses to invading agricultural insect pests is to attempt to lessen their impacts by using insecticides (Chapter 10.2). Unfortunately, insects are extraordinarily good at evolving resistance to these chemicals—more than 500 species have done this, often becoming simultaneously resistant to several different chemical groups. For example, the Old World diamondback moth, now a global pest of many crops in the mustard family, has evolved resistance to every chemical used against it as well as to the bacterium *Bacillus thuringiensis* (Bt). Sometimes resistance takes the form of a physiological mechanism that allows an insect to detoxify or to tolerate the chemical. Other times, resistance is conferred by an evolved behavior, which can be as simple as staying on the bottom of the leaf instead of the top. Because pesticides act as very strong agents of natural selection, it is not surprising that they should sometimes lead to very rapid evolution of resistance. This happens not only in insects. Recently, rabbits resistant to 1080, a main chemical used against them, have been found in southwestern Australia. Plants can also evolve resistance: in Florida, the Asian waterweed hydrilla has evolved a biotype resistant to the herbicide fluridone, the chief chemical weapon previously used against this species.

Behavioral evolution to resist pesticides is but one example of evolution of genetically determined behavior in invaders. Behavioral evolution is also exemplified by migration routes of introduced species. Chinook salmon are native to the Pacific Northwest but were introduced to New Zealand early in the 20th century. They have colonized several river systems along the east coast of the South Island, and these populations now differ genetically among themselves and from their North American source populations in several morphological and physiological traits related to different migration routes. The timing of spawning migrations into rivers differs by as much as seven weeks among the different river systems, and the different stocks also differ from each other in ways related to the distance from the ocean feeding grounds to the river breeding ground. Fish that have a longer distance to go upriver tend to be bigger and to have smaller ovary mass and smaller eggs. A similar case involves mid-Atlantic shad introduced to rivers in the Pacific Northwest in the late 19th century. In these Pacific rivers, the shad now show an increasing frequency of repeat spawning and greater numbers of eggs than in the source mid-Atlantic populations and demonstrate genetically determined migration routes.

5.3 How do introduced pathogens and native hosts coevolve?

Many introduced diseases cause huge death of native or introduced hosts (Chapter 3.5), after which the host evolves resistance to the disease. Actually, a coevolutionary interaction occurs. From the standpoint of the pathogen, it is often not good to be too virulent, because if the host dies too quickly, the pathogen might be unable to reproduce or disperse. On the other hand, if the virulence is not adequate to insure enough reproduction by the pathogen, natural selection should cause the pathogen to become more virulent. This reasoning leads to the notion of "optimal virulence."

Myxomatosis in rabbits (Chapter 3.5) is an excellent example of a disease evolving optimal virulence. The *Myxoma* virus, native to the Americas, causes only a mild disease in its original New World hosts. It is transmitted by flea or mosquito bites, depending on the site. Myxomatosis was introduced during the 1950s to many regions to reduce introduced rabbit populations, and initially the pathogen achieved this objective spectacularly, killing 99% of the rabbit populations in Europe and Australia. However, with the passage of time, rabbit populations have been less and less affected. In Australia, there have been five epizootic waves of disease, each killing a smaller proportion of rabbits. What is the origin of this change in virulence? It is certainly partly due to elevated resistance among the rabbits—those that manage to survive transmit to their descendants the genes that limited the viral impact. However, rabbit resistance does not entirely account for the phenomenon. Comparison of frozen stocks of the virus collected several decades apart shows that the virus nowadays is, on average, much less virulent than the virus that began the first epizootic. Why? The most virulent variants of the virus killed rabbits so rapidly that insect vectors did not have time to transmit them to new hosts before the originally infected hosts died. So natural selection acted against the genes that caused the virus to be most virulent. However, if the virus multiplies too slowly, the probability that the fleas or mosquitoes ingest virus in the course of their blood meals declines, which means that the genes of viral types whose virulence is too weak will also not be propagated. Natural selection maintains virulence at a compromise, intermediate level, at which viral invasion rate suffices for the fleas and mosquitoes to become contaminated while biting but does not cause the rabbits to die too quickly.

Chestnut blight in North America is yet another disease that can evolve differing degrees of virulence. It is caused by an Asian fungus transmitted from tree to tree by various insects in two strains—one virulent and one hypovirulent (less virulent). When the two strains compete on the same tree, the virulent strain outcompetes and eliminates the hypovirulent one, and the host tree dies quickly. At the scale of the chestnut population, this competitive outcome causes a mass decline of chestnut of the sort seen in North America in the first half of the 20th century (Chapter 3.1). However, when chestnuts become too rare, the advantage shifts. The less virulent strain becomes increasingly favored. What causes this competitive shift? The fact that the less virulent strain does not kill the tree too quickly means that insects are more likely to transmit it rather than the virulent strain to a new tree. Once again, this shows that virulence is a two-edged sword for a pathogen: if virulence decreases host reproductive success too much, natural selection for lower virulence ensues.

Evolution of increased virulence on the part of a pathogen or a parasite can come at some expense—it often cannot do other things as well because it has evolutionarily "invested" more in virulence. An interesting example of this phenomenon is seen in a parasitic braconid, introduced in 1957 from Europe to control alfalfa weevils in the western United States. The wasp was initially only weakly effective against the Egyptian alfalfa weevil (which had been introduced in 1939), because an immune response of the weevil larvae killed 35–40% of the wasp eggs. However, by 15 years later, the wasp had largely overcome the weevil's immune response, and egg mortality was only about 5%. However, the wasp had become correspondingly less effective than it had been against another species of weevil, the alfalfa weevil.

Sometimes the odds are stacked so heavily against a native host that an introduced pathogen is likely to overcome any resistance the host evolves. White pine blister rust, for instance, evolves virulent strains very rapidly. Native to Asia, this fungus arrived in eastern Europe in the 19th century in Asian trees introduced to botanical gardens. In Europe it attacked the eastern white pine—a species native to North America that had been widely planted in Europe. The fungus then got to North America in the early 20th century on imported white pine seedlings. It infects eastern white pines and related species by virtue of spores produced on alternate hosts—plants of the currant and gooseberry genus. Massive pine death led to a futile, expensive campaign to eradicate currants and gooseberries (Chapter 9.3). In Asia, pines resist the fungus, but those in North America have been devastated. There is some genetic resistance to white pine blister rust among native pines, but the rust has been able to evolve faster than the pines (not surprising, given the great disparity in generation length and population size between trees and fungi), so it easily overcomes resistance. Some previously resistant stands of trees like western white pine and sugar pine were eventually defeated by more virulent rust strains.

5.4 How do native species evolve in response to invasions?

Evolution of resistance is just one of several ways native species evolve in response to invaders. Native Asian birds, for instance, are highly resistant to the Asian avian malaria parasite (Chapter 3.5) that has so ravaged Hawaiian bird populations. Very recently, however, a single Hawaiian bird species has evolved resistance. The Oahu 'amakihi has become resistant in about 150 generations and now occurs with impunity in the lowlands, where other native birds are excluded by malaria. Another remarkable example of a bird species, in this case an invasive one, becoming resistant to a subsequent invader has occurred on Hispaniola. The village weaver was introduced from Africa in the late 18th century, probably as a cage bird. In its native Africa, this species is parasitized by cuckoos and shows high levels of egg rejection—that is, discriminating the eggs of parasitic cuckoos and throwing them from the nest. Until recently, no parasitic birds inhabited Hispaniola, and non–weaver eggs placed in weaver nests were rarely rejected. However, during the 20th century, the parasitic South American shiny cowbird invaded the West Indies. First observed on Hispaniola in 1972, it began to exploit village weavers as hosts. However, within 16 years, rejection rates of cowbird eggs by weavers increased from 14% to 83%. It is uncertain whether this behavioral change is genetically determined—that is, whether it constitutes evolution—or whether it is a learned response, but at least a genetic contribution is suspected.

An interesting example of rapid morphological evolution of a native species in response to an invader is seen in the native North American soapberry bug. In Florida, some soapberry bugs have colonized the introduced Chinese rain tree, which has smaller fruit than the native host, a balloon vine. To access the smaller fruits of the rain tree, soapberry bugs have evolved mouthparts that are about 30% shorter than those of bugs on the native host. Still more remarkably, in Australia a native relative of this insect, the Australian soapberry bug, has evolved mouthparts 5–10% longer since the introduction of a South American balloon vine (different from the species in Florida). These insects originally specialized on a native plant, the woolly rambutan, but in 30–40 years they have shifted to feed on the larger fruits of the introduced balloon vine. Other instances abound of rapid morphological evolution of native species in response to invaders. After the introduction of the predatory green crab to the northeastern coast of North America in the first half of the 20th century, the native dog whelk and flat periwinkle, mollusk prey species, evolved thicker shells. In the southeastern United States, fence lizards have evolved longer legs that aid their escape from the South American red imported fire ant, which invaded in the mid-20th century.

Many native species evolve behavioral responses to introduced predators. The fence lizard, in addition to its morphological evolution, has responded to fire ant invasion by evolving a novel behavioral response, a tail flick followed by rapid running away from an ant mound. Fence lizards from the same species collected from regions not invaded by the ant do not show this behavior and are quickly overwhelmed by attacking ants. On the West Coast, the introduction of bullfrogs from the eastern United States has caused both the Pacific tree frog and the red-legged frog to evolve avoidance behavior in the presence of bullfrog chemical cues, while the red-legged frog also reduces activity.

Just as invading insects have evolved behaviorally and physiologically to use new native host plants, so can native insects evolve to use introduced host plants. A classic case is that of the apple maggot fly, which is native to North America even though apples are not. This fly originally fed on hawthorn. However, long after apples were introduced to North America (in the early 17th century at the latest), a biotype evolved that subsists on apple rather than hawthorn. This apple biotype has evolved to emerge much earlier than the hawthorn biotype, timed to the earlier appearance of apples than hawthorn, and it gradually spread from its origin in Massachusetts in 1864 so that it now occupies the great majority of the range of the hawthorn biotype. Some specialists even consider these biotypes different species, although there is still a small amount of hybridization between them.

5.5 How does hybridization affect natives and invaders?

Hybridization has already been discussed in terms of its impact on native species (Chapter 3.6), particularly with respect to the threat of genetic extinction of uncommon native species in the face of massive hybridization with a much more common invader. In a few cases, such as that of American and European mink, the hybridization is not followed by direct genetic changes to either the invader or the native because the hybrids either die or are sterile, so they cannot backcross to either parental species and thereby transmit genes from one to the other. However, in several instances, a greater or lesser degree of backcrossing occurs, and the genetic changes that ensue constitute evolution (Chapter 3.6). The most dramatic such evolution is the formation of totally new species, such as common cordgrass and Welsh groundsel (Chapter 3.6).

Still other species have come into existence by hybridization among introduced species. For instance, three European species in the same genus were introduced to eastern Washington State in the early 20th century: meadow salsify, western salsify, and salsify. Hybrids among them were soon noted but were found to be sterile, as were the hybrids of native and introduced cordgrass in Britain. But by the mid-20th century, a series of hybridizations among the crosses, followed by chromosomal mutations similar to those in cordgrass, had produced two entirely new salsify species, each with twice the number of chromosomes of the parental species. There is even evidence that similar hybrids among the same parental species have independently, but by the same means, produced the same new species in distant regions, such as Arizona.

Additionally, hybridization between introduced and native populations of the same species, or between two different introduced populations, can lead to the evolution of various new adaptations. Such adaptations can produce a variety of impacts. For instance, common reed is a major weed of disturbed wetlands and even invades undisturbed sites in North America. It used to be considered an Old World species, but recent research revealed that it has resided in the Southwest for at least 40,000 years and on both the Atlantic and Pacific coasts for at least several thousand years—it may well be native. However, its range and abundance have increased dramatically in the last 150 years, a spread spurred by introduction of an Old World population of the same species, probably in soil ballast of ships in the early 19th century. The genes of this Old World population did not spread rapidly until the expansion of railways and roads in the late 19th century, but now this introduced genetic strain of common reed has dispersed widely, hybridized with native strains, and common reed has even spread to areas not formerly inhabited by the species. Common reed has thus become a much more invasive plant.

A similar introduction of individuals from the Old World—in this case, from several European sites—has transformed the native species reed canary grass from a harmless, not very prominent component of local North American floras to a widespread, deplored invader that threatens native wetland species and impedes wetland restoration. The introduced populations arrived beginning around 1850 for use as a forage crop and to revegetate shorelines. By the late 19th century, this grass was widespread and often dominant in many wetland habitats. There has been such extensive hybridization and gene flow among these introduced populations that there is now more genetic diversity in North America than in any one site in the native range, because the introductions arrived from disparate, often widely separated regions of Europe. Furthermore, the invasive properties of the species, which include many life cycle and morphological traits, are due to various combinations of genes from the introduced populations, no one of which would likely have been particularly problematic by itself. Together, they have produced a formidable wetland pest.

Introductions similarly explain the expanded range of the invasive green crab in North America. First recorded in the United States in 1817 in New York and southern Massachusetts, it spread over a century to the Gulf of Maine, but not until the 1980s did it reach Nova Scotia. Genetic analysis shows that the genetic lineages in the northern part of the crab's range derived from separate invasions by genotypes that can survive in colder water; these probably arrived in ballast water of ships using new routes.

Many other examples are now known in which multiple introductions from different parts of a species' native range have led to genetically diverse introduced populations. A burst of research following the discovery of new ways to detect genetic variation, beginning in the 1960s, has led to one of the most important scientific discoveries of the last century—the existence of vast amounts of previously unsuspected genetic variation in natural populations. Among the populations in which such genetic variation has been detected are introduced species. The brown anole and the multicolored lady beetle (Chapter 3) are species invasive in North America for which genetic tools have revealed multiple introductions, and in both species the spread and invasive characteristics may be at least partly due to evolution beginning with mixing of genes from different source populations. In both instances, genetic analysis also shows that subsequent invasions came not from the native regions of the invaders but from the more genetically diverse introduced populations. For the brown anole, the invaders in Hawaii and Taiwan arrived from Florida, as they have genes and genetic combinations found in Florida but not the native Cuba. Similarly, the Asian multicolored lady beetle, introduced several times to North America for nearly a century, did not begin to spread explosively until the late 1980s. It appears that a much more widespread invasion was triggered by an evolutionary event involving hybridization in an eastern North American population comprising genetically different individuals from both east Asia and west Asia. This hybridization produced more invasive types of beetles, and the North American population then exploded in numbers, also serving as a bridgehead for invasions of Europe and South America.

Cheatgrass, one of the most invasive plants in North America, has similarly arrived multiple times and from different regions of the Old World. It was introduced at least seven times in the West and twice in the East. Also, there have been at least two introductions of cheatgrass to the Canary Islands, one from the Iberian Peninsula and the other from Morocco, and two to Argentina, both from eastern Europe. Genetic analysis strongly suggests multiple introductions for other nonnative species—mouse-ear cress (a favorite study organism of plant evolutionists) in North America, wakame (an Asian kelp) in Europe and Australasia, zebra mussels in North America, and the spiny water flea, a damaging Old World crustacean zooplankton species in the Great Lakes.

As demonstrated in previous examples, hybridization between populations from different sources can render an invading species more invasive. For the multicolored lady beetle, such evolution is strongly suspected, but the particular traits that triggered the invasion are not yet known. For reed canary grass, various life history and morphological changes appear to have produced more invasive strains. Like many parts of the New World, the Caribbean island of Martinique has been invaded by the Malaysian trumpet snail. Genetic analysis shows that there are several invasions from different sources and that hybrids among these have become more invasive by virtue of changed life history features—especially larger size at birth and reduced fecundity.

5.6 What are the enemy release and EICA hypotheses?

In 2002, ecologists Ryan M. Keane and Michael J. Crawley, scanning the huge impacts of some plant invaders, suggested that the reason most of these species could wreak so much havoc was that they had left behind their natural enemies— especially herbivores, parasites, and pathogens but also competitors—and so are at a great advantage when competing with native plants. The natives, in this view, must expend a large fraction of their resources defending against their enemies (for instance, by producing chemicals that deter them), and they nevertheless suffer substantial mortality and loss of reproduction at their hands. The invaders are free from such costs and can devote all their resources to growth and reproduction. This enemy release hypothesis at first blush seems diametrically opposite to Elton's earlier idea of biotic resistance (Chapter 2.5), the notion that most invasions fail to establish, or, if they do establish, remain restricted and rather innocuous by virtue of the native natural enemies that oppose them, except in places like islands that have few such natural enemies. However, the two ideas can be seen as part of the same hypothesis when we recall that most introduced species do not become invasive (perhaps because of biotic resistance) but a few do (perhaps because of enemy release). That is, the apparent contradiction can be resolved if neither hypothesis is viewed as a general, overall rule or law that explains the fates and impacts of all introduced species. Many rules in ecology and evolution are not rules in the sense of the laws of thermodynamics but simply statements of patterns that are more or less dominant, although there are always exceptions.

In fact, a close examination of some introduced species that have achieved greater densities and have greater impacts in their introduced ranges than in their native ranges shows that, in certain cases, escape from the effects of natural enemies does seem to be key to their success. White campion, a small perennial European plant introduced to North America in the early 19th century, has spread widely and become highly invasive. It suffers much heavier attack from insects and snails in its native Europe than in North America, and a fungus and a seed-eating moth caterpillar that devastate white campion in Europe are absent in North America. Overall, plants of this species are 17 times more likely to be damaged by natural enemies in Europe than in the introduced range. In other cases, however, absence of natural enemies seems to have little to do with a major invasion. For instance, the spread of Brazilian pepper in Florida (Chapter 4.3) results not from absence of natural enemies but from a progressive change in the physical environment making it particularly suitable for this plant. The enemy release hypothesis is the foundational principle for the method of biological control to reduce the impact of invasions, which will be addressed in Chapter 10.3—that is, natural enemies keep invasive pest species under control in their native ranges, so one or more of these can be imported from the native range of any invasive pest to lower its population size. As will be discussed in Chapter 10.3, sometimes this method has worked well, and other times it has failed utterly. Probably this reflects the many and varied underlying causes for the impacts of different invaders. Also, in some instances native species can either attack an introduced species immediately upon its arrival or, as seen already, can evolve quite quickly to do so. Thus, there is little reason to think that the enemy release hypothesis should explain the trajectory of every invasion, even though it may explain some quite well.

The enemy release hypothesis, in turn, has spawned an evolutionary hypothesis. Termed the evolution of increased competitive ability (or EICA) hypothesis, it was proposed in 1995 by ecologists Bernd Blossey and Rolf Nötzold. Observing some well-known cases in which individuals of introduced plant populations are vastly larger (and often produce more seeds) than individuals of native populations of the same species, they suggested that the absence of natural enemies should cause introduced plants to evolve to devote less energy to defensive chemicals and structures and more to reproduction (often through increased size).

Certainly many botanists have been struck by the impressive size and vigor of introduced populations relative to those of their counterparts in their native ranges. In 1864, W. T. Locke Travers wrote from New Zealand to the great English botanist Joseph Hooker: "You would be surprised at the rapid spread of European and foreign plants in this country. All along the main lines of road through the plain the Knot-grass (*Polygonum aviculare*) grows most luxuriantly, the roots sometimes 2 feet in length, and the plants spreading over an area of from 4 to 5 feet in diameter." In England this plant rarely exceeds a foot across. However, these cases of greatly increased size do not constitute a general rule. Ecologist Christophe Thebaud and I examined two large data sets and found no overall tendency for plants to be larger in their introduced than in their native ranges. Some species are, and these may indeed prove to be larger by virtue of enemy release. For instance, European ragwort does devote more resources to growth and reproduction in New Zealand, Australia, and North America than in Europe and is less defended against some specialist insects that attack it in Europe but are absent in the introduced ranges. However, just as many plant species are smaller in their introduced than in their native ranges (perhaps because the physical environment is less suitable for them, but this also would have to be demonstrated in each suspected case).

5.7 How can the paradox of invasion genetics be resolved?

We should now reconsider the paradox of invasion genetics with which this chapter began—most invasions begin with small numbers of individuals, yet many have not fallen to inbreeding depression or genetic drift, the threats associated with small numbers. Part of the paradox may be resolved if inbreeding depression is not as universally debilitating as it is widely advertised to be, and some theoretical evidence suggests that its effects are most severe when a formerly large population is suddenly greatly reduced in size and less harmful when a population has always been small or declines gradually. Theory also suggests that, if a small population can expand quickly, the impact of genetic drift can be far lower than if the population remains small for many generations. Many invasive introduced populations do increase rapidly upon introduction. However, some invasions spread only after a long time lag (Chapter 4.3), so this theoretical finding cannot completely resolve the paradox.

Another possible resolution is that those widespread invasions we observe today that originated from introductions of just a few individuals are simply the survivors among many introductions, and they are a small minority—all the others failed. Both inbreeding depression and genetic drift act probabilistically, so one would expect a few survivors even if the theory is correct. In truth, we know very little about the causes of extinction of most failed invasions. After all, if a few individuals of a plant species or an insect species arrive in a new place and either fail to reproduce or dwindle quickly to extinction, how likely is it that anyone would even record their transient presence, much less study why they disappeared? However, a number of theoretical arguments suggest that for very small populations that die out fairly quickly, genetic factors are much less likely to be the reason than environmental ones or simply the chance vicissitudes of demography, such as the likelihood that all offspring in a generation will be of one sex only.

In fact, for at least some invasions started with few individuals, genetic variation is greatly reduced, and the invader thrives and spreads anyway. For instance, the starlings in North America, whose introduction as a "Shakespeare's bird" was outlined already, lack about 42% of the genes found in the native European populations, and starlings from all over North America tend to be quite similar genetically, unlike European starlings. The American starling population increased very rapidly after introduction, which should have lessened the effects of drift. Still, this is one of several striking cases in which relative genetic poverty does not appear to have hindered a massive invasion.

As we have seen in the previous examples, part of the paradox may be resolved by the fact that, for some invasive species, even though the number of individuals initially (and subsequently) introduced is small, additional introductions, especially from different source populations, can bring great genetic variation. This variation may suffice not only to stem the actions of inbreeding depression and genetic drift but also even to allow the evolution of new, better adapted types. The flood of new genetic research showing previously unsuspected multiple introductions challenges the conventional wisdom that newly established populations of nonnative species founded by small numbers of individuals are automatically genetically impoverished. Such research has just begun, and its results so far are impressive; however, it is still too early to tell if multiple introductions play a major role in resolving the genetic paradox.

6

HOW AND WHY DO
INVASIONS OCCUR?

6.1 Why do people deliberately introduce animals?

Motives for introducing nonnative species are many. The underlying incentive behind all of these introductions, however, boils down to one simple truth: people are never happy with the species they have. This one phrase could serve as the motto for the entire field of invasion biology. Topping the list of motives for introduction is the desire for new food items. Few species introduced as food become invasive, but some do. Among animals, the giant African snail was originally introduced to Pacific islands by the Japanese during World War II as a potential food item. Both this mollusk and the predator introduced in an unsuccessful attempt to control it wrought havoc (Chapter 2.4). Asian immigrants brought the northern snakehead to the United States as a culinary delicacy (Chapter 1.1). This large, voracious, predatory fish from Asia was released in 2002 to a pond in Maryland by a man who had purchased two in a New York City market. Later the snakehead reached the Potomac River, and it is now established in the Maryland–Washington, D.C., area. Another marine species, the Asiatic clam (Chapter 3.6), probably arrived initially in North America as a food item, as did the Chinese mitten crab in California. In Africa, the Nile perch was introduced to Lake Victoria in an attempt to generate a fishing industry, with disastrous impacts on the native fishes (Chapter 1.3). What is worse, wealthy European exporters dominate the Nile perch fishery, and native fishermen were devastated economically and socially. The North American signal crayfish and red swamp crayfish (Chapters 2.3 and 3.5) were introduced to Europe as food items. Colonists or passing sailors introduced rabbits to hundreds of islands worldwide to be used as food, with devastating effects (Chapters 2.5 and 3.4). Early explorers similarly introduced European wild boar to many islands as bush meat for subsequent sailors or colonists, and early Melanesians and Polynesians introduced a smaller pig variety for food as they colonized Pacific islands. These pig species have been one of the most devastating island invaders, causing many of the same impacts as rabbits. Finally, the ravages of reindeer introduced to subantarctic islands to generate food for whalers were noted in Chapter 2.5. In sum, even though only a small fraction of all animal species introduced as food items have become damaging invaders, those that have include several disastrous cases, particularly in aquatic habitats and on islands.

Other animal species have been introduced for sport hunting and even more for sport fishing, although sometimes the line between introduction for food and introduction for sport is not easy to discern. Among game birds, various quail and pheasant species, ducks such as the mallard, and several pigeon and dove species have been widely introduced for hunting. Some such as the pheasant, gray partridge, and chukar partridge in North America and the mallard on several islands (Chapter 3.6) have become numerous. Of course, people also eat these game birds, but the key impetus for their introduction was for hunting, not for food. Himalayan tahr (a relative of goats) were introduced to New Zealand a century ago for sport hunting and have become an invasive threat to native plant communities. Sitka black-tailed deer were introduced in the late 19th and early 20th centuries as game animals to Haida Gwaii (the Queen Charlotte Islands) on the west coast of Canada. The deer have completely changed the plant community of the islands by browsing on preferred tree species and ground cover plants, with follow-on effects on many animal species. The raccoon dog, a feared Siberian carnivore, was introduced to Europe as a game animal and furbearer. It eats not only mammals, birds, frogs, and lizards but also prized wild berries, and in addition it is a vector of rabies. In Denmark they are so feared that hunters are asked to shoot them on sight.

The number of fish species introduced for sport fishing worldwide is staggering. Lago Nahuel Huapi, in the Patagonian Andes of Argentina, now draws fishermen from all over the world—but the fish they prize are North American rainbow and brook trout and European brown trout. Meanwhile, the native fish species have become far less common and restricted to the edges of the lake. Introduction of game animals and sport fish has often been undertaken by government agencies. For instance, the State of Florida has long maintained a laboratory whose raison d'être is to introduce new sport fish to the state. Private individuals also move game animals and especially sport fish to new locations, sometimes in rogue introductions that deliberately contravene government efforts to limit or eradicate invaders. Perhaps the most famous such case concerned the northern pike, introduced to Lake Davis in northern California. The California Department of Fish and Game, fearing that this large predator would move downstream and devastate salmon and trout populations, mounted an effort in 1997 to eradicate the pike by poisoning the lake. The effort failed, possibly because of reintroduction by fishermen who prize the large, challenging pike.

Many other species have been introduced as bait, often simply by casual release by fishermen. For instance, the sheepshead minnow, a nonnative bait fish, was released in such numbers in the Pecos River in Texas that it produced a hybrid swarm with the native Pecos pupfish over much of the range of the latter species. The barred tiger salamander, distributed as bait for sport fishing, has been introduced into the range of the native California tiger sálamander, threatening the native species with extinction by hybridization. The upper Midwest has no native earthworms; all species present were introduced after the retreat of glaciers beginning about 18,000 years ago. Now fishermen discarding bait have widely distributed two European worms, red wrigglers and nightcrawlers, which have wide-ranging effects on ecosystems (for instance, greatly thinning litter layers) in forests where native species are adapted to a wormless environment. Additionally, bait packaging can carry invaders. For example, green crab juveniles abound in kelp used to pack live worms.

A number of animal species have been introduced as food for commercial or sport fish and occasionally for game animals. The opossum shrimp introduced to boost kokanee populations, with subsequent disastrous results (Chapter 4.1), is but one example. A related shrimp species was introduced to Scandinavian lakes to provide food for sport fish, triggering a decline in populations of native water fleas. In Haida Gwaii (the Queen Charlotte Islands, Canada), North American red squirrels were introduced as prey for martens, a mammal prized by trappers. In one of the strangest motivations for introduction, government officials also released red squirrels in the hope (never realized) that they would gather spruce cones and seeds used in commercial forestry. The squirrels established populations on several islands, and though they have apparently increased marten populations they also prey on songbird nests, contributing to a decline in bird populations caused by the impact of introduced deer on vegetation.

Many mammals have been introduced as furbearers, notably South American nutria to Great Britain and other European countries as well as parts of the United States, North American muskrat to Europe, and foxes to several Alaskan islands. The aquatic nutria undercuts banks and causes land to be lost to swamps or open water; the muskrat has similar impacts. Arctic foxes devastate seabird colonies, with subsequent massive vegetation changes as nutrient cycling is greatly modified by loss of guano. One of the most destructive introduced furbearer invasions is that of the beaver to Tierra del Fuego (Chapter 3.1).

Innumerable animals kept as pets escape or are deliberately released when owners tire of them. Among pets that have become invasive are the huge Burmese python in Florida (Chapter 3.3) as well as its more aggressive relative, the African rock python, also in Florida. More mundane pet species can also be enormously destructive when they become feral. Feral housecats kill millions of native songbirds each year, and roaming dogs threaten native mammal and reptile populations. Escaped or released cage birds have often established populations, some of which have become invasive. For example, the rose-ringed parakeet from Africa and Asia and the South American monk parakeet have both established invasive populations in Europe and North America. Both of these parakeets threaten crops, and there is concern that the rose-ringed parakeet will compete with native birds for nest sites. Colonies of the monk parakeet damage electrical utility structures, in which they build large communal nests. In Europe, released pet red-eared slider turtles from Florida compete with native turtles and consume tadpoles that avoid native turtles but not the invader. Among other potentially invasive species kept as pets are crayfish, ants, a Chinese mantis, and the giant Madagascan hissing cockroach. The variety of species coveted as pets is astounding, reflecting the breadth of people's tastes as well as the desire to have the weirdest pet on the block.

One of the more bizarre rationales for introducing a species was advanced by a Hungarian entomologist. In 1951, George Bornemissza arrived in Australia and quickly noticed the staggering number of cowpats on the landscape, in contrast to the situation in his native Hungary. He conceived of the government-backed Dung Beetle Project, which introduced approximately 50 dung beetle species from Africa and Europe into Australia between 1964 and 1985. The idea behind the project was that the dung beetles would bury livestock feces. Dung beetles eat dung of herbivores and omnivores, either burying it in place or rolling it in balls up to ten times their own weight and then burying it in a more propitious location. The beetles also lay their eggs in buried dung, so their hatching larvae have an immediate food source. The massive introduction of sheep and cattle into Australia presented a problem, because native dung beetles had adapted to the very different type of dung excreted by kangaroos and other marsupials. Twenty-three of the species introduced by the Dung Beetle Project flourished and largely removed the piles of dung. Meanwhile, native dung beetles in Texas do bury cattle dung, but not quickly enough for some ranchers. In the 1970s, their displeasure resulted in the release of a dung beetle native to Africa and Asia through a program of the United States Department of Agriculture (USDA). The beetle has now spread widely in Texas. A related European dung beetle species was introduced to the Florida Panhandle around 1970 and has now dispersed throughout Georgia and as far west as Mississippi.

Species are often introduced through biological control projects—that is, deliberately introducing natural enemies of targeted introduced pest species (Chapter 10.3). Many established introduced insects in many parts of the world were originally introduced for biological control. The United States has at least 150 wasp species that were purposefully introduced as biological control agents for insect pests. In Florida, of about 1,000 established nonnative insect species, about 50 were introduced for biological control. Previously discussed invaders that were originally introduced for biological control of another pest include the small Indian mongoose, weasel, stoat, ferret, grass carp, mosquito fish, cane toad, rosy wolf snail, seven-spot lady beetle, multicolored lady beetle, tachinid fly, Australian parasitic wasp, South American parasitic wasp, and the bacterium *Bacillus thuringiensis* (Bt).

6.2 Why do people introduce plants?

As with animals, few of the myriad plant species introduced for food have become invasive. Of seven crop plants listed as major by the USDA, none are native; most are from the Old World. Yet none of these nonnative species invade natural habitats. Perhaps the most important invader among plants introduced for food is strawberry guava, brought to Hawaii in the early 19th century both for food and as an ornamental. This species has spread widely into forests and become a significant threat to local forest biodiversity. Aficionados of strawberry guava and products such as jam made from it have actively opposed a biological control program targeting this species. In Brazil, Asian jackfruit has become invasive. For instance, in the Tijuca Forest National Park near Rio de Janeiro, jackfruit planted in the 19th century has recently become problematic. Small mammals such as common marmosets (themselves introduced in this region) and coatis eat the copious seeds of the huge fruits that crack open when they fall to the ground. These animals disperse the seeds, which produce seedlings and saplings that compete with native plants. In addition, the mammals prey on eggs and nestlings of native birds. Park authorities have culled thousands of jackfruit saplings in an attempt to limit the damage.

Many plants have been introduced or distributed as feed for livestock, and some have become invasive. Often the same plants have been used with greater or lesser success to control erosion. For example, Johnson grass, from the Mediterranean region, was long distributed as a forage grass for livestock and for erosion control in the United States until it was recognized as a costly weed, particularly in cotton and soy fields. Kudzu, Russian olive, and multiflora rose, all from Asia, were originally imported and distributed by the USDA

Natural Resources Conservation Service to slow erosion, and all have become major environmental weeds. Kudzu has also been planted as a forage plant. In Australia, buffel grass from Africa and Asia and para grass from Africa were both introduced as livestock forage and became highly invasive. The former species was also introduced to Texas and Mexico for the same purpose and was additionally distributed for erosion control; it is now regarded as one of the worst invaders in the Sonoran Desert, particularly in Arizona. African molasses grass was similarly introduced to Hawaii as a forage grass and is now a feared invader that fosters forest conversion to grassland through increased fire frequency (Chapter 3.1). South American pampas grass has been distributed in the United States as a possible forage crop and for erosion control and has become a detested invader. In Brazil, South African lovegrass was introduced as a forage crop but turned out to be too fibrous. Cattle avoid it, and it now outcompetes native species in an area of over five million acres.

Nonnative plants are also sometimes introduced as wildlife habitat. For instance, Amur honeysuckle was brought to the United States as an ornamental but subsequently was widely distributed by the Natural Resources Conservation Service to improve bird habitat. Ironically, recent evidence shows that introduced honeysuckles actually decrease reproduction of some native bird species, and Amur honeysuckle has become an ecologically damaging invader.

Some plants have been introduced for fiber, such as cotton for cloth and trees for paper. Some tree species have subsequently invaded native ecosystems, especially conifers such as Monterey pine from California in Chile and South Africa, Douglas fir from the Pacific Northwest in Argentina and New Zealand, North American lodgepole pine in New Zealand and Chile, North American loblolly and slash pine in Brazil, and maritime pine from the Mediterranean region in South Africa. Several species of Australian acacia are invasive in South Africa. Plants other than trees have also been introduced as building materials. For instance, Asian giant reed, which has transformed riparian areas and changed fire cycles, was introduced to California as a roofing material. Recently, several introduced plant species have been cultivated as biofuel feedstock, including such highly invasive species as reed canary grass and giant reed. It is worrisome that the rapid growth and hardy nature that would help make a plant species a promising source of biofuel might well render it highly invasive.

Plants have also been introduced to effect landform changes. On Oahu, Floridian red mangrove and Asian large-leafed mangrove were brought to an intertidal wetland in the early 20th century with the goal of converting it to dry land. These species, especially red mangrove, have spread to other intertidal sites around Oahu and other Hawaiian islands, creating forests where none had previously existed. Mangrove roots accumulate sediment, shed tons of leaves per year per acre, and have completely transformed the habitat over large areas. In a similar case, Australian eucalyptus species were brought to Israel in the early 20th century to dry up swampy areas and render them suitable for agriculture, an effort that, in concert with other actions, has largely succeeded, although eucalyptus in Israel has not spread widely beyond the areas in which it was planted.

Deliberately introduced horticultural varieties are the plant analog to animal pets, and in many regions plants deliberately introduced as ornamentals comprise at least half of all highly invasive plant species. The practice of bringing striking plants from afar began around 1600 in Europe and a century later in the United States. Famous figures introduced some of the most feared invaders: Thomas Jefferson likely introduced Scotch broom to North America, and the American landscape architect Frederick Law Olmsted brought Japanese knotweed. Among important aquatic invaders brought to North America as ornamentals are South American water hyacinth and giant salvinia, Brazilian waterweed, and Eurasian watermilfoil. During the 19th century, seed catalogs in the United States routinely offered species that subsequently became major invaders, such as Brazilian pepper and Johnson grass.

6.3 What were acclimatization societies?

In the 18th and especially the 19th centuries, as biological exploration of new regions burgeoned, it became fashionable to establish acclimatization societies. These were social clubs (some with government encouragement) whose goal was to introduce species from afar, especially those deemed "useful" in some way, such as birds likely to kill insects or that have cheerful songs and pretty plumage. Such societies thrived in the United States, Great Britain, France, and other countries and also on remote islands colonized by their citizens, like the Hawaiian Islands or the Île de la Réunion in the Indian Ocean. Hawaii even had two competing acclimatization societies: the Hui Manu, dominated by Anglophones; and the Honolulu Mejiro Club, founded by Japanese immigrants and dedicated to introducing birds from Asia. The American Acclimatization Society was active in the United States, releasing starlings, English skylarks, pheasants, chaffinches, blackbirds, English titmice, and English robins with varying degrees of successful establishment. One of its members was Eugene Schieffelin, whose goal was to introduce all birds mentioned by Shakespeare; he succeeded with the starling (Chapter 5.1). On Tahiti, Eastham Guild operated as a one-man acclimatization society, introducing no fewer than 35 bird species between 1938 and 1940. Only two of these species established populations: the silvereye from Australia and New Zealand and the crimson-backed tanager from South America.

6.4 What are some unusual motives for introducing animals?

In east and Southeast Asia, the practice of releasing prayer animals for religious purposes is widespread, among Christians as well as members of eastern religions. These prayer animals frequently include introduced species, and the scale of such releases is often enormous. For instance, in Taiwan, 29.5% of a random sample of citizens reported releasing prayer animals, purchased from pet stores that, in turn, buy them from dealers or trappers. A popular prayer animal is the pond slider turtle from Brazil, which is now the second most abundant turtle on Taiwan. Several exotic fish species are also released in prayer ceremonies. Among birds commonly released in Taiwan are the common myna, Javan myna, and jungle myna from other parts of Asia as well as the Chinese bulbul, which is threatening the native Styan's bulbul with extinction through hybridization.

A persistent concern is that invasive species will be deployed as a form of ecoterrorism. During both World War II and the Cold War, the United States was charged with dropping Colorado potato beetles into Europe to decimate potato crops, although the evidence is ambiguous. On a much smaller scale, a disgruntled laborer introduced the small Indian mongoose to the small Caribbean island of Fajou to spite a landowner with whom he had a dispute over pay—he expected the predator to kill chickens and ducks belonging to his nemesis. One of the more bizarre motives for introducing species is scientific recognition. On the Scottish island of Rum, a prominent English botanist fraudulently bolstered his academic reputation by covertly introducing several plant species, then announcing their "discovery."

6.5 What are steppingstones?

Often species are deliberately moved long distances to sites where they are not highly problematic but from which they can move on their own, or as human-assisted stowaways, to nearby regions and become highly invasive. The initial colonization sites are called *steppingstones*. For example, the collared dove of Asia and Europe was brought to the Bahamas and Lesser Antilles as a cage bird, but individuals that either escaped or were deliberately released produced a population that dispersed to Florida by the 1980s. Their arrival was followed by a dramatic spread throughout much of the United States, reaching the Pacific coast by the late 1990s. These birds are now a noisy pest and may be competing with native birds. The South American cactus moth was introduced to the small Caribbean island of Nevis to control pest cactus. From this steppingstone, it island-hopped all the way to the Greater Antilles. From there it reached Florida either under its own volition from Cuba or in cut flowers transported from the Dominican Republic. Spreading north from Florida to South Carolina and west to Mississippi, it now threatens native cactus species in the United States and Mexico. In retrospect, one can see that the original deliberate introduction to Nevis was bound to spread farther to sites where the species was unwanted.

6.6 How do unintended introductions occur?

Although many of the most damaging invasions are unintended consequences of deliberate introductions, species have often invaded by hitchhiking on human conveyances, stowing away in various cargos, or otherwise taking advantage of human activities to get to distant new sites. Certain pathways of introduction have brought many invaders; other species have arrived by more idiosyncratic means.

As discussed earlier, ship ballast often carries species inadvertently to new homes. Many of the first introduced insects recorded in North America were beetle species typical of the ground and soil of Cornwall, a reflection of the fact that ships bound to North America to pick up timber and other cargo would stop at Land's End to load soil ballast for the trip across the Atlantic. In modern times, ballast water is a pathway for hundreds of aquatic invasions. High-impact invaders previously mentioned that probably arrived in ballast water include the zebra mussel, sea walnut, *Beroe ovata*, and spiny water flea.

The green crab invasion of the Atlantic coast of North America probably began with soil ballast, while the subsequent invasion of the Pacific coast is attributed to ballast water. Ship hulls fouled by encrusting organisms also transport many species to new homes. For instance, a study of introduced marine animals in the North Sea concluded that about 30 such species had arrived on fouled hulls.

Many invasive insects and plant pathogens have hitched rides in untreated timber or wooden packing. The emerald ash borer and the Asian longhorn beetle are two invading forest pests of North America that recently entered in this way. The former species has already devastated many stands of ash, while national, state, and local agencies have fought a 10-year battle to keep localized invasions of the latter species in the East and Midwest from spreading. Beetles that spread Dutch elm disease are believed to have hitchhiked to North America on untreated elm logs, while chestnut blight arrived either on chestnut timber or live saplings. More recently, the chestnut gall wasp reached Italy on chestnut cultivars from China between 1995 and 1996, the North American western yellow jacket probably reached Hawaii on untreated Christmas trees, and the Old World sirex woodwasp likely reached the New World and South Africa in untreated timber. Once established in a new region, such species spread locally, often by hitching rides with unsuspecting humans, as when campers bring firewood into forests.

Ornamental plants are often invasive in their own right, but they can also carry hitchhikers such as insects and mites that can establish and become invasive. For instance, the Japanese beetle entered the United States in iris bulbs. In Florida, the Mexican bromeliad weevil arrived in nonnative ornamental bromeliads and attacked native bromeliads, while the Central American cycad weevil hitchhiked on exotic cycads and attacked native ones. Balls of soil surrounding transported nursery plants are also a prime pathway for hitchhiking insects, snails, pathogens, and other species. The New Zealand flatworm, a carnivorous species devastating earthworms in parts of Great Britain (Chapter 2.5), arrived by this means. Cut flowers can also carry many hitchhiking invaders, including mites and thrips. Seeds of desired plants are often contaminated with seeds of other species, some of which are highly invasive. Cheatgrass, for instance, was probably introduced to North America as a contaminant of crop seed. The degree of contamination can be staggering. An experiment in the United States in which scientists ordered aquatic plants from nurseries found that 93% of the shipments had unwanted plant or animal contaminants in addition to the desired plants, and 10% of the contaminants were species prohibited for sale and shipment federally or by state regulation.

Similarly, introduced oysters are often themselves invasive, particularly when they produce complex surfaces in soft-sediment locations, as the Pacific oyster from Asia has done in North America and Europe. But oyster culture is also a source of hitchhiking invaders. Two damaging oyster diseases were introduced with the oysters—dermo and MSX (Chapter 3.5). Other invaders arrive as contaminants of oyster shipments or as part of packing material for such shipments. For instance, an Asian turf-forming red alga was introduced with the Pacific oyster and forms massive intertidal monocultures in the Azores and California; it is also an aquarium pest. Wireweed, a highly invasive Asian alga, was also introduced to the West Coast with Asian oysters, as was Japanese eelgrass, which has transformed mudflats in Washington into seagrass meadows. Smooth cordgrass from the East Coast of North America was introduced from the East Coast to the West Coast when it was used as packing for eastern oysters being shipped to Washington State by rail. It has massively transformed mudflats to grass beds, much as common cordgrass has in Great Britain (Chapter 3.6). Worse, smooth cordgrass is now hybridizing with a native California cordgrass in California, and the hybrid is invasive in much the same way as common cordgrass is in Great Britain. The Japanese oyster drill, a predatory snail, was introduced to the West Coast of North America with oyster seed from Japan, while the Atlantic oyster drill was introduced to the West Coast with introduced eastern oyster seed. Both drills attack native mollusks. The common slipper shell from the North American East Coast hitchhiked to the West Coast and Europe on shipments of eastern oyster; it competes with other filter feeders and has become a commercial problem in Europe. Although most hitchhikers on oysters are either damaging or neutral, occasionally one is beneficial, at least to humans. The eastern soft-shell clam, for instance, became a valued food item after accidental introduction to the West Coast with introduced eastern oysters. It is not known to cause any kind of harm.

Escapes from fur farms are yet another common source of unplanned invasion. North American gray squirrel and mink and South American nutria in Great Britain, North American raccoon and muskrat in parts of Europe, and North American beaver in Tierra del Fuego are among species that have invaded by this route. Myriad more idiosyncratic pathways have been important routes for particular groups of invasive species. Paving stones and ceramics, for instance, carry many hitchhiking mollusks. The gypsy moth invasion of North America began in a misguided experiment to establish a silk industry in the United States; in 1868 moths escaped from the insectary in which they were kept, in Medford, Massachusetts. Gypsy moths spread in North America by attaching their egg masses to the undersides of motor vehicles, while introduced brown tree snakes from Guam have reached Hawaii (but have not yet established populations) by stowing away in wheel wells of airplanes. Road building and farm machinery often transport unwanted plant seeds and soil animals. Whirling disease of trout arrived in the United States in frozen rainbow trout, after rainbow trout had been introduced to Europe and acquired the parasite there (Chapter 3.5).

On June 7, 2012, a floating dock 66 feet long, 19 feet wide, and 7 feet deep and weighing 188 tons, ripped from its moorings in Japan by a tsunami more than a year earlier, washed

Figure 6.1 End view of Japanese dock that washed ashore in Oregon after being loosened by a tsunami. (Photograph: Courtesy of Oregon State University Coastal Oregon Marine Experimental Station.)

ashore in Oregon (Figure 6.1). It carried at least 104 nonnative marine species embedded in over 2 tons of seaweeds and encrusting animals. Two days later, a team from the Oregon Department of Fish and Wildlife wielding scythes and blowtorches tried to eradicate the various mussels, crabs, barnacles, worms, kelps, and other species before they established populations. The outcome of the operation remains to be seen. This was just one of four similar docks torn free by the tsunami. A second one was photographed drifting by the Hawaiian Islands in September 2012 but has not been seen since. As this book goes to press, a third one has just washed ashore in a remote part of the Olympic Peninsula, Washington, over 20 months after the tsunami, and a team of federal and state officials and scientists is rushing to the site.

7

CAN WE PREDICT SPECIES INVASIONS?

7.1 Can we predict which species will become invasive if they are introduced?

Asa Gray, the great 19th-century American botanist, contended that impacts of invasions cannot be predicted, and invasion biologists from Charles Elton onward have lamented how hard it is to foresee what a newly introduced species will do. However, many are now more optimistic than Gray was. The fate of any introduction has an element of uncertainty partly because two universal features of living organisms are fundamentally unpredictable. First, all living organisms, including introduced species, either move on their own or have adaptations that cause them to be transported by wind, water, or other organisms. Second, all living organisms reproduce, and they evolve over a number of generations. Both movement and evolution are heavily influenced by chance. In addition, all living organisms interact with other species. Introduced species interact with native species and often with other introduced species in bewildering myriad ways (Chapters 3–5). Certain sorts of interactions and outcomes can be readily predicted— if we introduce a predator like the ship rat or the small Indian mongoose to an island naturally lacking such predators and containing ground-nesting bird species, we can be quite certain the birds will suffer massive mortality, and possibly even extinction. However, many other interactions could not easily have been deduced in advance. Who would have guessed that introducing opossum shrimp to the Flathead catchment would cause a bald eagle population to crash (Chapter 4.1)? Or that introduction of the rabbit disease myxomatosis to Great Britain would lead to extirpation of a butterfly species (Chapter 4.1)?

Attempts to predict which introductions would lead to damaging invasions have taken one of two tacks. First is to try to use traits of species, such as size, means of dispersal, or rate of reproduction, to predict which species would be invasive. This approach builds on previous studies of weeds and is captured in the notion of a list of traits that make an ideal weed, as suggested in 1965 by Herbert Baker, an American botanist. The Baker list contains traits such as long-lived seeds that germinate in a variety of environmental circumstances, rapid growth, high rates of seed production, seeds adapted to both short- and long-distance dispersal, and self-compatibility (that is, pollen can fertilize ova of the same plant to produce seeds). One can imagine that such traits would indeed be very useful for an individual plant arriving in a new place and "wanting" to initiate a new invasion. However, analysts soon realized that many invasive plants lack some of the key traits on the Baker list, while other plants having most or all of the traits were not invasive. This finding led invasion biologists for a time to deemphasize species traits in their attempts to predict invasion impacts. However, several studies in the 1990s suggested that if the comparison of species is restricted to species within groups of rather similar species—say, pine species or woody plants—species traits might yield quite accurate predictions. For example, American ecologist Marcel Rejmánek and South African invasion biologist David Richardson found that three traits in particular distinguished pine species that had become invasive from those that had not: seed size; interval between large seed crops; and minimum length of the juvenile period before saplings start producing seeds.

However, even within groups of similar species, species traits occasionally fail to predict correctly which species will become invasive. Several pine species that have been predicted to be highly invasive based on their traits and that have been invasive in many regions have failed to invade a Patagonian site in spite of massive plantings. Many other species have been introduced to multiple sites with different results in different locations, including rainbow trout, bluegills, and climbing asparagus. It is therefore clear that traits of species are not the only determinants of whether species become invasive once introduced to new sites.

The other common approach to predicting whether a newly arrived species will become invasive is the biotic resistance hypothesis (Chapter 2.5). This is the idea that any introduced species is countered in its new home by an array of native species that oppose or resist it—competitors, predators, parasites, and pathogens. The biotic resistance hypothesis is supported by the observation that two kinds of sites—islands and habitats disturbed by humans (such as road verges)—are often heavily invaded, and both of these habitats have relatively low numbers of native species. The reasoning is that this impoverished biota provides less resistance to any newcomer, so introductions commonly lead to invasions. However, it is apparent that, even though there is a pattern of more invasions in certain types of sites and even though that pattern results at least partly from lack of biotic resistance, this pattern does not suffice to allow good predictions about the fates of particular introductions. This inability becomes obvious when one considers that even sites with very many species—such as some prairies and forests—do experience some invasions, so at least some introduced species do surmount whatever resistance the native species pose. Also, some introductions fail to become established and to spread even on species-poor sites, such as islands. Something else determines success or failure of an invasion more often than simply the biotic characteristics, such as number of species, of the invaded site.

In short, the trajectory of any introduction—whether or not an invasion will ensue—is determined by a combination of factors, including traits of the introduced species itself and features of the site to which it has been introduced.

Two other aspects of invasions complicate predictions, the frequent time lags associated with invasions (Chapter 4.3) and the reasons, including invasional meltdown (Chapter 4.2), that such lags terminate. As discussed earlier, the reasons for such lags, and for their ending, are sometimes mysterious. Other times it is the arrival of a subsequently introduced species that turns an innocuous species into an invasive problem. A good example is the spread of a fig in Florida only after the arrival of the fig wasp that must pollinate it. Whatever the factors that cause a quiescent resident to become an invader, they reflect contingency—that is, one phenomenon (an invasion) is contingent on the occurrence of another, independent phenomenon. In general, the outcomes of any phenomena (not only species introductions) that consist of chains of contingent events are extremely difficult to predict, as there is often no good reason to consider the triggering event that constitutes each step in the chain. Consider the sequence of events by which the introduction of Chinese grass carp in Arkansas threatened a native fish in Utah, the woundfin. First the carp needed to arrive with an Asian tapeworm. Next the tapeworm had to infect native fishes, and one of those had to be a popular baitfish that would be carried westward to the Colorado River. From the Colorado River, infected fish would have to reach the part of Utah with the native fish species. It would have required a form of clairvoyance for someone considering possible impacts of the Chinese grass carp in North America to have concocted this unlikely sequence of events as a possible route to a harmful impact.

7.2 What is risk assessment for biological invasions?

Despite the risks of a harmful impact of planned introductions, many people desire to introduce species: as crops, pets, and ornamentals; for biological control purposes; as biofuel feedstocks, and for many other purposes. At the same time, the growing recognition of the harm invasions can cause has led regulators and scientists to try to develop various formal tools to try to estimate and quantify the risk of harmful impacts. In fact, many multilateral treaties, such as those of the World Trade Organization (WTO) and the North American Free Trade Agreement (NAFTA), mandate that nations provide a quantitative risk assessment when they seek to impose barriers on import of living organisms.

Among risk assessment tools currently employed, probably the one most widely used for permitting purposes is a method called the Australian Weed Risk Assessment (AWRA), devised for use with proposed plant introductions in 1999 and currently in regulatory use in Australia and New Zealand. It consists of 49 questions about the history, range, and biology of species proposed for introduction. The answer to each question is awarded between −3 and +5 points, depending on the extent to which the answer suggests a risk of invasion. The AWRA score is the sum of points for all questions. Depending on this score, there are three possible outcomes: accept the species for importation; reject the species; or demand further evaluation for invasive potential. The cutoff scores for the three outcomes are arbitrarily set, according to the degree of risk the regulators are willing to accept. In tests applying the AWRA to introduced species that are already known to be major invaders in Australia, New Zealand, Florida, and several other locations, with cutoff points determined by the investigators, the AWRA successfully identified between 82% and 100% of these invaders and would have rejected them for introduction. However, the AWRA would also have rejected a few species that have not turned out to be invasive. A study of the actual application of the AWRA to 2,800 plant species proposed for introduction through 2006 revealed that 53% were accepted, 27% were rejected, and further study was demanded for 20% of these species. Recent research by invasion biologists Curtis C. Daehler and John G. Virtue suggests that predictions of the AWRA can be improved by segregating questions relating to likelihood that a plant species will be introduced from those concerned with consequences if it establishes.

In another risk assessment method, a series of experts consider all the various risks they can think of that might accompany a planned introduction. For instance, the United Kingdom requires that any proposed introduction be subjected to the UK Non-Native Organism Risk Assessment. British regulatory authorities use this assessment as an aid in deciding what introductions to permit. They ask experts to assess, for each introduction, both the likelihood of an identified risk and the uncertainty each expert feels in estimating that likelihood, using a five-class scale for each. The various combinations of likelihood and uncertainty are then arbitrarily combined in a matrix (Table 7.1) that shows whether

Table 7.1 Risk acceptability matrix used in the UK risk assessment scheme for nonnative species

	Uncertainty				
Likelihood	Minimal	Minor	Moderate	Major	Massive
Very unlikely	*Negligible*	*Negligible*	Justifiable	Justifiable	Justifiable
Unlikely		*Negligible*	Justifiable	Justifiable	Justifiable	Justifiable
Possible	Justifiable	Justifiable	Justifiable	Justifiable	**Unacceptable**
Likely	Justifiable	Justifiable	Justifiable	**Unacceptable**	**Unacceptable**
Very Likely	Justifiable	Justifiable	**Unacceptable**	**Unacceptable**	**Unacceptable**

Notes: The outcomes yielding negligible risk are in italics, and those yielding unacceptable risks are in bold.

Source: Adapted from R.H.A. Baker et al. 2008. The UK risk assessment scheme for all non-native species. *Neobiota* 7:46–57.

there is negligible risk, some risk but one that is justifiable, or an unacceptable risk. The assessments of different experts must then be combined (for example, by averaging them) to give an overall assessment. Of course, whether or not one would want to approve a release would rest on more than just an estimate of the likelihood that there would be some impact. It would also depend on the size of the impact, which might be ecological, economic, or of some other sort. For example, there may be a very small risk (say 1%) of a certain sort of impact, but if that impact occurs, it is devastating in some way (say, causes the extinction of a species or destruction of an ecosystem or crop). One might then be far less inclined to permit the introduction than if the same 1% probability refers to an impact that is itself small (for example, a slight reduction in some native population or a slight loss to some crop). Some risk assessment procedures define risk in terms of both magnitude and probability by arbitrarily combining them in various ways. New Zealand's Environmental Risk Management Authority combines them to yield four risk categories: insignificant, low, medium, and high (Table 7.2). Again, some arbitrary formula must be established for averaging the scores of individual assessors.

Table 7.2 Classification of risks associated with a planned species introduction with both likelihood and magnitude of an impact taken into account, as tallied by the New Zealand Environmental Risk Management Authority.

	Magnitude or Cost				
Likelihood	Minimal	Minor	Moderate	Major	Massive
Very unlikely	Insignificant	Insignificant	Low	Medium	Medium
Unlikely	Insignificant	Low	Low	Medium	High
Possible	Low	Low	Medium	Medium	High
Likely	Low	Low	Medium	High	High
Very likely	Medium	Medium	High	High	High

A general problem with risk assessments of the previous sort, resting on expert judgment, is that different experts may assess *magnitude* from very different perspectives. Therefore, averaging the scores of several experts might not adequately capture the entire gamut of views on the magnitude of the possible impact of some introduction. Such different views often arise in consideration of release of biological control agents (see Chapter 10.3). In these cases, a farmer may see the potential benefit as high (for example, saving a crop from an insect pest or weed) and the risk low, while a conservationist may see the potential benefit as relatively unimportant but the risk as high (for example, the possibility that the agent may attack a threatened nontarget native species). Thus, although risk assessments can force people to discuss many different sorts or risks and can yield quantitative results, they are nevertheless subjective, both because the expert panel members may have different values and because the algorithms combining probability, uncertainty, and magnitude or cost are arbitrary.

A component of some risk assessments, and often a first attempt at predicting the impact of a nonnative species that has already established a population but has not spread widely, is a class of models known as species distribution models (SDMs). These models are computer-based tools that record physical environmental conditions (broadly speaking, the species' niche) in its native range, and sometimes in the part of the new region that it already occupies, and try to use this information to predict where the species will ultimately reach as it invades. Will a South American insect arriving and establishing in Miami be likely to spread to North Florida? To the rest of the South? To all of North America? Such modeling can be used as a "quick and dirty" means of risk assessment before more biological and observational information becomes available to inform policymakers and manager about courses of action. Should they try to act quickly to eradicate an incipient invasion, even if the attempt would be costly and success not guaranteed? Should they ignore the invasion because the species is unlikely to be able to survive except where it was first established? Sometimes the answer seems obvious even without examining data. A palm species that happens to be established in a continuously heated area (say, next to the exhaust of a laundry) in Maine or a tropical fish population established in a heated pond in Minnesota would be unlikely to spread unless carried a great distance by humans, deliberately or inadvertently. Other times, the answer is not so evident.

Consider the Burmese python, now widely established in South Florida. Will it spread north to, say, South Carolina? West to Louisiana? A team of federal biologists attempted to answer this question by considering the climate of 149 known locations of this python species in its native Asian range and constructing what is know as a *climatic envelope* based on temperature and precipitation that aimed to define the region of the United States that has climate similar, at least in terms of temperature and precipitation, to that in the native locations. Their conclusion was that a wide swath of the United States, encompassing the South and spreading north to Delaware and west to California, could potentially be inhabited by this snake if it could reach these places. This prediction was contested by other herpetologists, largely on the grounds that only certain parts of the species' Asian range are occupied by the race or subspecies of pythons found in Florida. They believe the possible extent of the invasion is far more limited. In 2010, a record-breaking freeze in Florida killed many pythons, but not nearly all, and those in captivity in the northernmost part of their introduced range were unable to attempt to bury themselves and hibernate, as they might have in nature, because the cages in which they were kept prevented their burrowing. Thus, this dramatic event did not clear up the controversy.

The climatic envelope method used for the python was necessarily simplistic because climatic records from much of the native range are scant and the biology of the python is not well-known. A widely used, more sophisticated but related method, known as CLIMEX, attempts to model the impact of climate on population growth and survival of the introduced species. A bewildering array of other statistical methods is now available, often with readily accessible computer programs, to predict the ultimate range of an introduced species from climatic and other physical factor data from its native range. These methods often have intriguing acronyms and abbreviations for names: GARP, GAM, GLM, BRT, ROC. The best method for a particular introduction depends largely on the specific data available. Many methods require that the data from the native range be systematically gathered, through surveying a carefully selected set of sites in the native range. Probably the most popular current approach—MaxEnt—makes the fewest assumptions about the data in the native range, so it is appropriate when surveys have been sporadic and haphazard (as usually is the case for data readily available from field guides, for example). MaxEnt has been employed to predict the ultimate ranges of dozens of introduced species, but recent statistical analysis suggests that a newer procedure, MaxLike, provides more accurate estimates.

Predictions generated by all species distribution models are subject to criticism on several grounds. Perhaps most important is that they assume that species ranges will be determined by climate and other physical factors and do not take account of species interactions. Yet it is indisputable that some species' ranges are absolutely limited by the presence of another species (for example, a predator or pathogen). It has been suggested, for example, that lake trout were unable to cross to Europe and Asia on their own because of predation by large lampreys in both the Atlantic and Pacific Oceans. The fact that they thrived once introduced to Europe and Asia supports this hypothesis. In several cases of invasional meltdown (Chapter 4.2), introduced species expanded their range only upon introduction of other species. A method that does not take account of the presence of other species can make very inaccurate predictions if interactions are important in determining a species' range. SDMs also assume that introduced populations will not evolve, yet we know they do (Chapter 5). Evolution could include adaptations to somewhat different climatic conditions. Such evolution must occur frequently, as natural selection would automatically operate to favor those individuals best adapted to the newly colonized environment. What is more at issue is the speed at which such evolution will occur. In some cases, it has occurred quite rapidly. For instance, Eurasian barnyard grass has recently expanded its introduced North American range northward into colder climates in Quebec, an expansion that has been tied to physiological evolution—a change in certain enzymes.

Because of ongoing global climate change, an innovation that should become increasingly common in use of species distribution models is to apply them not only to the climate in the potential introduced range as it is today but also to the climate as it is predicted to be several decades or a century in the future. Several models predicting global climate change—called general circulation models (GCMs)—are now widely used not only in climatology but also in many other fields. Species distribution models often show a huge difference in predicted range for a minor difference of 2 or 3 degrees Fahrenheit, and GCM models are generally in agreement that mean temperature in many regions will change by at least that much over the next century. So it is only reasonable to consider future climates when using species distribution models to predict geographic ranges of introduced species. Most invasion biologists believe that global climate change is likely to exacerbate problems associated with invasions, and this belief is largely grounded in predictions that some introductions that now fail will succeed in a warmer world and others that are already established will spread (Chapter 12.3).

An alternative to species distribution models to predict the eventual range of a newly established introduced species is mechanistic modeling. In this approach, careful measurements of various tolerances (for instance, maximum and minimum temperatures for survival) are made in the laboratory and then are compared to field conditions in the new home of the introduced species. Such modeling of course entails more time than application of a computer model to available literature data on climate and species locations, and it is subject to the same two criticisms just articulated—it would not take into account either species interactions or evolution. It is thus not nearly as frequently used. However, as this is a more direct way of measuring immediate response to different climatic conditions, it is probably more accurate than species distribution models, unless the latter are based on a very large number of sites in the native range of a species. A mechanistic model used in 1979 predicted that Asian itchgrass could spread from Florida and Louisiana to the states of the Gulf Coast, lower Midwest, southern Atlantic coast, and Southwest. So far the plant has reached the entire Gulf Coast, parts of the southern Atlantic coast, and Indiana.

7.3 How is risk assessed for unplanned introductions?

As hard as accurate risk assessment is for deliberate introductions, it is vastly more difficult for inadvertent ones. The problem is that all of the previously listed tools—which species traits might conduce to invasiveness, which climates an introduction might tolerate, and the like—pertain to particular species, and if we do not know which species are going to arrive, we cannot apply these approaches. However, another form of risk assessment has been applied to pathways. In this method, a pathway is assessed to see if it is likely to introduce one or more species that may become invasive.

For instance, the United States Department of Agriculture (USDA) Forest Service conducted a risk assessment for importation of untreated Siberian larch logs into the United States. The concern was not whether Siberian larch itself would become invasive but whether forest pests or pathogens might inadvertently be transported on the logs. The team broke the invasion process into four stages, each initially to be assessed independently: (1) the probability of pest organisms being on or in the logs to be transported; (2) the probability of pests surviving the voyage; (3) the probability of pests establishing once they land in the United States; and (4) the probability of pests spreading beyond the point of establishment. The team then attempted to assess the magnitude of possible economic and ecological impacts if a pest were to spread. A key problem with the process was that there were too few data on most of the 175 known pest species that inhabit Siberian larch in its native range—for example, insects, fungi, nematodes. The team thus chose 36 species that they hoped would represent the three means of transport—hitchhiking on logs, inhabiting bark, and inhabiting wood—to attempt to estimate the risk of an invasion. For each of these 36 species, the team members estimated probabilities for the previous four stages: (1) probability of being on Siberian larch in the region of origin; (2) probability of actually being carried on a transported log and surviving the transport; (3) probability of establishing a population after arrival; and (4) probability that the established population will spread. Step 2 was divided into two substeps: (1) being on a log; and (2) surviving the trip. For each of these probabilities, the possible rankings were low, low-medium, medium, medium-high, and high. For each stage, the individual rankings were averaged to give a team ranking. As these stages are a sequence of independent events, the total probability of an invasion by each species would be the product calculated by multiplying its stage probabilities. Unfortunately, presumably because of the wide uncertainty associated with most entries, the team did not proceed beyond classifying the probabilities for each stage. Thus, for example, for five bark beetles among the species assessed, the results were as in Table 7.3.

The team used just six species (three insects, a nematode, and two pathogens) to attempt to estimate the ecological and economic consequences if an invasion occurred. Independently, these six pests were predicted to have an economic impact of anywhere from $25 million to $58 billion over 50 years, but the team did not attempt to say how accurate these estimates were likely to be, nor did they attempt to combine them. For ecological impacts, the team simply listed the potential impacts of each of the six species but did not attempt to quantify or judge the likelihood of each impact.

Table 7.3 Probabilities of passing each stage of invasion for five bark beetles that may infest Siberian larch brought to North America

	Probability of being on larch	Probability of being on transported log	Probability of surviving transport	Probability of establishing population	Probability of spreading
Engraver beetles	H	H	M	M/H	H
European spruce bark beetle	L	H	H	L	H
Great spruce bark beetle	M/H	H	H	H	H
Conifer weevils	M	H	H	L	H

Notes: L = low, M = medium, H = high.

Source: From U.S.D.A. Forest Service Miscellaneous Publication 1495, 1991.

One immediately spots the difficulty of such a risk assessment. The procedure consists of a series of guesses. The guesses may be educated guesses if the assessors are experts, as was the case for the Siberian larch risk assessment; however, they are still guesses. An accurate quantified statement, such as "There is a probability X of a risk of magnitude Y," is simply impossible. The real value of such a pathway risk assessment is in the formal attempt to consider all the possible risks that the pathway generates.

8

HOW ARE SPECIES
INTRODUCTIONS REGULATED?

8.1 What international agreements address biological invasions?

Most biological invasions arise from the deliberate or accidental transport of species from one nation to another, so it is no
surprise that, as the enormous impacts of invasions came to be
recognized in the late 1980s (Chapter 1.2), a number of international treaties and agreements began to address such transport. Perhaps most all-encompassing, at least in principle, is
article 8h of the Convention on Biological Diversity (CBD, or
the "Rio Convention") convened in 1993: "as far as possible
and as appropriate," each signatory nation shall "prevent the
introduction of, control or eradicate those alien species which
threaten ecosystems, habitats or species." Alone among major
nations, the United States has not ratified the CBD, which now
has 193 nations as parties. Another problem is that the phrase
"as far as possible and as appropriate" can be variously inter-
preted, and individual nations that are party to the convention draft national laws and regulations that vary greatly in
their interpretations. Nevertheless, the CBD is a major sym-
bolic international document enjoining nations to prevent,
manage, and control biological invasions. The CBD has held
10 meetings of member states since Rio (the 11th is scheduled

for 2012), and these have provided more specific guidelines and targets.

Even before the CBD, an International Plant Protection Convention (IPPC) in 1951 formulated a series of phytosanitary measures to keep plant pathogens and pests from spreading in international commerce in plants; this convention was revised in 2005. Several other international conventions deal with specific aspects of biological invasions. For instance, in 2005, the International Maritime Organization adopted guidelines to implement practices proposed at a convention the previous year on management of ship ballast water and sediments to prevent the spread of living organisms. Regional international conventions also exist that contain substantial sections on biological invasions, such as the Convention for the Protection of Natural Resources and Environment of the South Pacific Region (1986) and the Convention on the Conservation of European Wildlife and Natural Resources (1979).

A general problem with international regulations generated by these and other conventions and treaties is that individual nations are left to implement the directives. Some conventions can impose penalties for failure to regulate invasions, but this seems not to have happened. On the contrary, multilateral treaties can be used to hinder national efforts to regulate invasions. Treaties of both the World Trade Organization (WTO) and the North American Free Trade Agreement (NAFTA) require that any attempt by a signatory nation to exclude a shipment from another nation (for instance, a living organism or a product that might carry a living organism) must be accompanied by a quantitative risk assessment validating the claim that an economically or environmentally harmful invasion might ensue. Otherwise, the exclusion is viewed as a form of economic protectionism masquerading as an environmental concern. In Chapter 7, we saw that, unfortunately, accurate quantitative risk assessments for invasions are currently extremely difficult if not impossible, so this requirement is difficult or impossible for a nation to fulfill. In an important test case, Australia tried to exclude the shipment from Canada of frozen salmon on the grounds that these could carry parasites that would infest native Australian fish. This is a very valid concern, in light of the fact that whirling disease of trout entered North America in exactly this way (Chapter 3.5), as Australia pointed out. Yet, when the Canadian government appealed the exclusion, the WTO sustained the appeal, ruling that Australia had failed to produce a *quantitative* risk assessment and was also being capricious, as ornamental fish were not excluded on the same grounds. The Australians were thus forced either to accept the frozen salmon or to suffer a sizeable financial penalty.

8.2 How do national regulatory frameworks differ?

New Zealand is viewed as the gold standard in regulating introduction of nonnative species by virtue of two parliamentary acts, the Biosecurity Act of 1993 and the Hazardous Substances and New Organisms Act of 1996, which established a strong framework for keeping out potential invaders and for responding rapidly when an incipient invasion is detected. One key feature is that no species whatsoever can be deliberately imported without prior consideration by an expert committee concerning invasive potential and also likelihood of carrying invasive hitchhikers. There are no blanket exceptions, such as pets or crop plants. Once a committee has determined that threat is minimal, the species may be put on a white list of approved species. The committee may also put the species on a black list of species whose import is strictly prohibited. Or it may delay a decision pending further study, relegating the species to a gray list. Border security in New Zealand is also extraordinarily tight. Airports and seaports must be approved as places of first arrival, and all planes, ships, and boats must give notice of their arrival and prevent certain goods from leaving the vessel without permission from an inspector. All passengers and goods are subject to inspection. Surveillance and monitoring are also strong in New Zealand, and several species have been found and eradicated before they had spread widely, such as the Australian painted apple moth in the Auckland region. Finally, publicity regarding the threats of introduced species and warnings of penalties for violating regulations are omnipresent in New Zealand, beginning with extensive notices posted at airports and seaports.

Nevertheless, there have been breaches of biosecurity, most notably a rogue introduction of a calicivirus that causes rabbit hemorrhagic disease. The virus was in the testing stage in Australia as a biological control agent for European rabbits; however, infected rabbits escaped, and the virus established in Australia. New Zealand farmers illegally imported and spread the virus. Another problem in New Zealand is that far more resources have been devoted to preventing and managing pests and diseases that threaten agriculture, trade, and human health than to preventing those that threaten native biodiversity and the environment. The New Zealand parliament is currently considering revisions of the Biosecurity Act that would make it more effective and less costly.

Australia, like New Zealand, has a long history of damaging invasions, such as those of the European rabbit and red fox, South American prickly pear, and North American cane toad (Chapter 2). Problems caused by these invasions have led to an aggressive approach to introductions. Australia, however, is 30 times larger than New Zealand, leading to considerably greater difficulties in intercepting nonnative species and monitoring for established introductions. The great size of Australia also means that species native to one part of Australia may be just as nonnative and potentially invasive in another part as if they had come from a foreign country. Thus, some regulation and management occurs at the national level (the *federation*), some at the state level, and some at the local level. In particular, the island state of Tasmania and the state of Western Australia, which is separated from the rest of the country by a huge desert, have worked especially hard to prevent invasions. Tasmania strictly forbids import, even from other Australian states, of certain blacklisted species, while Western Australia deals similarly with plants and also has a long-running program to prevent Eurasian starlings, established in eastern Australia, from gaining a foothold in the west. Movement between states is regulated by the states with different species designated for exclusion by each state.

Australia and its states rely heavily on risk assessment for estimating both the possibility that a species proposed for introduction will become invasive and the possibility that it will carry a hitchhiking pest or pathogen. The Australian Weed Risk Assessment (Chapter 7.2) is an example of the former type. Notably, Australia requires a risk assessment if a species is already in the country but a proposed import is from a different source. Risk assessment results in a species being placed on a black list, white list, or gray list, as in New Zealand. Australia also uses both risk assessment and cost–benefit analysis when a nonnative species is found to have breached the interception barrier and established a population. The likely cost of an invasion and the feasibility of eradication are quickly estimated in an emergency procedure. As does New Zealand, Australia heavily publicizes invasive species problems as well as regulations regarding introduction and movement of nonnative species.

An isolated island nation, even a very large one like Australia, has an enormous advantage in trying to keep out invaders and in preventing reinvasion following eradication. This advantage is the absence of borders with neighboring nations that have different regulations and degrees of commitment regarding nonnative species. The United States, similar in size to Australia but sharing long borders with Mexico and Canada and having the West Indies very nearby, faces greater challenges. Full recognition of the scope of that challenge came in 1993 in the form of a congressional report lamenting the ad hoc nature of American responses to invasions and absence of adequate national laws. In fact, the United States relies primarily on two old black list laws. The Lacey Act of 1900 allows the federal government to forbid entry to designated animals, but very few species have been added to the black list, almost all of them after the species had already invaded (such as the house sparrow, starling, zebra mussel, and Chinese mitten crab). Worse, the Fish and Wildlife Service, charged with studying suggested additions to the Lacey Act list, takes on average 4 years to complete each study and has made final determinations in only 19 cases in the last 50 years. Many well-known invasive species whose impacts have been verified by scientific research are not listed under the Lacey Act, such as the giant African snail (Chapter 3.3), the green crab (Chapter 5.5), and the common myna (Chapter 2.4). The upshot is that federal, state, and local governments spend millions of dollars every year trying to manage damaging invasive species like the huge Nile monitor lizard in Florida while citizens can simply walk into pet stores or go online to buy them. In 2012 the Secretary of the Interior issued an emergency ban on importing or shipping across state lines four species of giant constrictors, including the Burmese python (Chapter 3.2), African rock pythons, and yellow anaconda, years after the first signs of impact from the Burmese python were recognized. Aside from the routine sluggishness of moving an action through a large bureaucracy, this ban was delayed by extensive lobbying by snake enthusiasts and pet dealers. One congressman even argued that banning such snakes would bring financial ruin to thousands of mom-and-pop pet dealers.

For plants, the analogous black list to the Lacey Act is the Federal Noxious Weed Act of 1974, superseded by the Plant Protection Act of 2000, which contains approximately 100 species, most of which are already present in the United States.

Congress has also passed a series of laws with much more limited focus, aimed at zebra mussel, salt cedar, brown tree snake, nutria, as well as other species, often in specific regions. However, Congress has not yet effectively addressed the general issue of invasive species—the approach has been largely ad hoc and politically motivated.

A sea change in national attitude was marked by President Bill Clinton's 1999 Executive Order 13112, which acknowledged that the federal effort had thus far been weak and established a National Invasive Species Council of cabinet-level representatives, charged with establishing a National Invasive Species Management Plan. The plan was developed in consultation with many stakeholders and was approved by the council in 2001. However, the plan was vague, and its goals have largely been unmet. It codified existing roles of various federal departments, such as the United States Department of Agriculture (USDA) and the U.S. Department of the Interior, despite the explicit recognition by the Executive Order that past efforts had been ineffective. A revised plan in 2008 has not progressed beyond the original version in important ways. However, under new regulations known as Q37, a gray list category like New Zealand's has been added to the blacklists currently in force.

The various executive branch agencies in the United States issue regulations if a federal law exists that appears to call for such regulation. The USDA, the U.S. Fish and Wildlife Service, and the United States Coast Guard are particularly active in formulating and implementing invasive species regulations. However, agencies cannot regulate if no federal statute mandates regulation. Thus, the Fish and Wildlife Service sought to establish a white list approach for introduced vertebrates but was denied this authority in the absence of laws calling for such regulation. By contrast, agencies sometimes do not regulate in situations where they would seem to have legal authority. The United States Environmental Protection Agency has authority under the Clean Water Act of 1972 to regulate ballast water but did not act until courts ordered it to do so in 2008 as the result of a lawsuit. Federal agencies often work with state agencies in formulating and implementing regulations. For example, state agencies largely implement the Clean Water Act, and various state agencies interact with the Department of Agriculture on quarantine regulations for pests of agriculture and forestry.

As with Australia, the sheer size of the United States means that the flora and fauna in any particular place have evolved in isolation from many species native to other parts of the United States. Species introduced from one part to another part of the United States can therefore become as highly damaging an invader as a species arriving from a foreign country. The impact of smooth cordgrass from the East Coast on West Coast ecosystems (Chapter 6.6) and that of the barred tiger salamander on the California tiger salamander (Chapter 6.1) are examples (Figure 8.1). States have attempted with varying degrees of success to stem such introductions, particularly Hawaii, which regulates such imports and publicizes the dangers to agriculture and the environment, and California, which especially focuses on potential pests of agriculture. Several states, especially in the Northeast, have lists of invasive plant species that are banned for transport and sale within the state. In addition, several states have tried to prevent the deliberate import from other states of certain foreign species already invasive or judged likely to become invasive. This effort has proven difficult because of the absence of clear-cut legal authority to do so in many cases as well as the Interstate Commerce Clause of the United States Constitution, which severely limits the ability of one state to prevent movement of goods (which can include living organisms) from one state to another. The Interstate Commerce Clause also makes it difficult for states to prevent hitchhiking organisms from arriving from other states on various commercial goods. States have also attempted to prevent other states from introducing foreign species on the grounds that

Figure 8.1 (L) Adult hybrid between California tiger salamander and barred tiger salamander. (R) Adult barred tiger salamander. (Photograph courtesy Jarrett Johnson.)

such introductions would likely lead to invasion not only of the state pondering the introduction but also of neighboring states, but such efforts have failed. Notable examples are the attempt of neighboring states and Canadian provinces to keep the North Dakota Game and Fish Commission from introducing European zander and objections by Missouri to the introduction of grass carp into Arkansas.

The United States, with only two bordering nations, has a far easier time regulating and managing introductions than European nations do. Europe has no comprehensive framework for dealing with invasive species. In 2011, the European Commission (the executive branch of the European Union) recognized that previous policies on biological invasions were ineffective and set out to produce a new set of regulations by 2012 that would identify and constrict pathways of introduction as well as prioritize particular invasive species for eradication or management. The regulations are to cover prevention (including movement within as well as between member states), early detection and rapid response, eradication where possible, and management where eradication fails or is not attempted. It remains to be seen whether this stated commitment translates into effects on the ground. Many of these intended activities are already supposed to be carried out in one form or another under existing European Union policy. Furthermore, as in the United States, action on invasions has often been reactive rather than proactive. Thus, for example, existing European Union policy regulates the trade and import of species that are already present and invasive in Europe, such as the red-eared slider, ruddy duck, and bullfrog, much as the Lacey Act and Plant Protection Act in the United States forbid import of species that are already present and invasive.

The general European Union problem is that the 27 member states have different regulations and, so far, differing degrees of commitment to regulating movement and acting on invaders. For instance, 15 of the states do not include certain large groups of species when listing which species are invasive. Member states use different sorts of risk assessments, and some members (e.g., Bulgaria, the Czech Republic, and Slovakia) do not use formal risk assessment at all. Yet failure of one state to monitor and act quickly on an invasion puts neighboring states at great risk. What is worse, some European nations are not part of the European Union and thus are ineligible for certain types of European Union funding available to member states. An example of the problems this situation can cause is the spread of the small Indian mongoose, introduced beginning in 1910 to control venomous vipers on islands that are now part of Croatia. This highly invasive species affects reptiles and amphibians in Croatia, and it has recently spread to the mainland of Croatia as well as to Serbia, Bosnia, and Montenegro. It is likely to disperse further into Europe, either on its own or by deliberate or inadvertent human transport, but the invaded countries are relatively poor and are uninterested in attempting to eradicate or impede the spread of the mongoose. Because they are not members of the European

Union, these nations do not qualify for funds that member states could use for this purpose, and the European Union seems unaware of the major threat it faces. But even if the European Union were alerted to the danger, there appears to be no mechanism for the Union to force a non–member state to act in such a case or to act on behalf of that state.

8.3 Do further introductions matter once a species has invaded?

Most people assume that, once a species is established, further introductions of the same species cannot exacerbate any problems the invader may cause. This view is codified in many laws and regulations on movement of organisms, especially plants. Washington State, for instance, assigns weeds to three categories. Class A are newcomers that are still rare; class B are those that are well-established in some counties but rare or absent in others; class C are common throughout the state. Landowners are obligated by law to attempt to eradicate all species in class A as well as those in class B in areas where they are not yet established. However, they need not do anything about those in class C, as this is viewed as a wasted effort—the species is already established.

However, the advent of molecular tools has shown that, even in well-established introduced populations, further introductions can enhance invasiveness or even render a harmless species invasive by bringing in new genotypes. In the cases of common reed, reed canarygrass, and the green crab in North America, such evolution has been confirmed (Chapter 5.5), while it is strongly suspected in the spread of the brown anole in Florida and the multicolored lady beetle in Europe (Chapter 5.5). It is apparent that laws and regulations regarding both planned and unplanned introductions badly need reconsideration in light of these findings. For instance, for planned imports of species that have already been introduced, the permitting process should at least consider what the source population is for the new import, and, if it differs from the source of the established population, proceed with great caution. As observed already, Australian procedures already do this. For unplanned introductions, pathways that might bring individuals from novel source areas might be particularly stringently monitored or regulated.

A related issue that demands improved regulation in light of experience concerns imports of native species from foreign sites where they have been introduced. The general thinking is that, since these species are native (even though the individuals are not), they will not become invasive, though concern that they will carry parasites can lead to the imposition of quarantines—that is, the requirement that they be held in isolation to ensure that they are healthy. Yet two pathogen invasions of North America—whirling disease of trout (Chapter 3.5) and white pine blister rust (Chapter 5.3)—arrived exactly this way. In fact, if one wanted to introduce new pathogens of some native species, one could hardly find a more effective method than growing that native species in a foreign country and then reintroducing it (or products made from it) back to its native range! The fact that many pathogens are not detected in quarantine suggests that any proposal to bring living or dead individuals of a species back into its native range should be viewed with extreme caution.

8.4 How could economic measures aid regulation and management?

From a market standpoint, movement of species around the globe as currently regulated is irrational in the sense that invasions and attendant damage are viewed as externalities of trade. That is, they are a consequence of market transactions but are not taken into account by whoever is engaged in the transactions. If damage occurs from a subsequent invasion, someone else (usually society as a whole or some large group of stakeholders) ends up bearing the cost, not the importer or exporter. So long as someone exporting or importing species either does not know about their presence and potential damage or knows but is not legally responsible for any subsequent damage, there is nothing to dissuade him or her from engaging in the activity. The current regulatory frameworks either permit or forbid an importation, period (except for a few cases in which further study is demanded). Once an importation is permitted, neither the importer nor the exporter is liable for any ecological or economic harm that may ensue. What is required to make the system more rational is to internalize these costs.

One way to do this would be to impose a tariff on trade in living organisms or on pathways likely to carry them. Tariffs in general are discouraged by international trade treaties such as those of the World Trade Organization because they are seen simply as a form of economic protectionism that hinders the operation of the free market. However, nothing is sacred about WTO treaties, which are occasionally renegotiated. Free market principles would not be invoked if one nation forbade the entry of a harmful chemical pollutant or radioactive waste, and the damage to the environment, the economy, and public health caused by biological invasions is at least the same order of magnitude as that imposed by chemicals (Chapter 4). It is a matter of recognizing the size of the risk and its probabilistic nature. So long as we insist on viewing every proposed deliberate introduction or every pathway in isolation and subject it to a risk assessment of the sort described in Chapter 7.2, it would be difficult to argue against many introductions on economic grounds alone. However, if the principle is accepted that a certain fraction of introductions will become problematic but we cannot tell very well which ones, then it would make sense to impose a tariff on all introductions to internalize what is now an externalized cost. A similar sort of tariff could be imposed on goods shipped by particular pathways, perhaps adjusted to correspond with the frequency of harmful invasions known to have arrived by each pathway.

Policy analyst Peter Jenkins has suggested a variety of other methods that could internalize what are now externalities of invasions. The main ones are:

(a) An importer of a living organism would be required to buy insurance against possible future damages if a population of that organism becomes a damaging invader.

(b) An importer of living organisms would have to be bonded in order to be certain that funds would be available should the need to eradicate or manage a damaging invasion be caused by one of his or her imports.

(c) Civil fines could be imposed on an importer if an introduced species becomes a damaging invader.

(d) Criminal penalties or fines could be levied if an introduced species became harmfully invasive.

(e) Fees could be imposed if an importer is introducing a species.

(f) Taxes could be imposed on importers as a class.

Methods (a) through (d) are reactive: costs are paid only after an invasion has occurred. This is particularly problematic because of the occasional lags that occur between introduction and invasion (Chapter 4.3). Such lags mean that the major expense to the importer or the insurer or bonding agency might be imposed many years, even decades, after the import. The importer or his or her agents might no longer exist, making it questionable that the expense would ever be charged to them rather than to society. Another problem is that the cost of a particular invasion (e.g., the zebra mussel) may turn out to be so great (Chapter 4) that even an insurance company or bonding agency could not cover it and the importer would be unable to pay a fine; the expense would again by borne by society as a whole. Thus, the more effective approaches would be (e) and (f), which are proactive. Pathways would be obvious loci for the fees envisioned in method (e). For instance, an importer receiving a plant (including seeds) or animal that is alive could be charged a fee, or the operator of a ship or plane that lands after an intercontinental voyage could be charged a fee. Fees or taxes could be routed by legislation toward activities that would ameliorate invasion problems as a class—for example, inspection procedures and management and eradication campaigns (Chapter 9).

9

DETECTION AND ERADICATION OF INTRODUCED SPECIES

9.1 What is an early detection and rapid response system?

Invasion biology shares a philosophy with medicine—it is far better to prevent a harmful event or disease than it is to rely on the hope that some prescribed cure will be successful. Still, nonnative species arrive in any nation, no matter what the regulatory framework and no matter how many dollars are poured into barricades. If populations of such species establish, the next line of defense is an early warning system associated with a mechanism to ensure rapid response if invasion is likely or has already begun. In addition to signaling the presence of an unintended introduction, such a system of early detection and rapid response (EDRR) would ideally monitor for unexpected invasions of species such as horticultural plants or ornamental fish that were deliberately introduced either because they were not anticipated to become invasive or because no regulation forbade their import. An attempt at eradication is far more likely to succeed if a population is targeted early, before it has spread widely. It is also possible that some invasions, even if they cannot ultimately be prevented, can be greatly slowed by various management procedures, such as quarantines on moving materials on which the invader might hitchhike from an invaded to an uninvaded area.

The urgency of establishing an EDRR system in the United States became apparent in the late 1980s with the mounting costs of dealing with widespread plant invasions such as those of witchweed in the Carolinas. A national herbarium survey in 1991 was undertaken to find records of infestations of plants on the Federal Noxious Weed list, and several small infestations (usually less than 4 hectares) thus identified were successfully eradicated, including small broomrape and tropical soda apple at several sites in the South and wild red rice in the Everglades. These are all established invaders in the United States, but eradicating isolated populations can keep them out of a region for years. Development of a more comprehensive EDRR program in the United States began in the mid-1990s, as federal agencies started to coordinate detection with state and regional partners to attempt to locate invaders quickly. One of the most effective partners is the Invasive Plant Atlas of New England, a cooperative effort among the University of Connecticut, a private wildflower organization, and federal agencies. This program trained over 700 agency staff members and volunteers to find infestations of 100 invasive plant species in New England. Field reports are verified by project experts and entered into a database—over 10,000 findings have been recorded. These have been cross-referenced and combined with herbarium records, and the resulting data are used to determine whether a particular invasion should be managed and, if so, by what means. A long-term goal of many invasion managers and scientists is to develop a network of similar regional plant atlases that together would comprise a national invasive plant atlas. No such comprehensive system yet exists to facilitate early detection of animal invaders. Nor is there a unified rapid response mechanism for plant or animal invasions that are detected in the United States.

Even if a formal rapid response mechanism did exist, finding an incipient invasion in the form of a small population would not likely automatically trigger an immediate eradication campaign, if only because resources are limited. Rather, there would be some sort of risk assessment for invasion and impact as well as a feasibility study to see if an eradication attempt would be likely to succeed. The problem here is that, because some introduced populations spread quickly, the window of opportunity for eradication may close rapidly, and costs of such a campaign may escalate. Two successful eradications are discussed later in the chapter, one in Australia and the other in the United States, of species believed to be highly invasive. In each case, the success was possible because of very quick action (in one case, nine days after discovery). Several instances are also described of massive invasions that occurred after early detection because no action was taken. How to reconcile the possible need for rapid action with the limited resources available for management is a major challenge to any EDRR system.

Much of the Australian effort against invasions takes place at the state level (Chapter 8.2), and Australian states have various EDDR mechanisms. For instance, Victoria has a Weed Alert Program that relies heavily on 2,000 volunteer weed spotters who try to find infestations of plants that are not yet widely established in the state. They are trained to recognize over 100 such species and to consult with the National Plant Herbarium of Victoria on questionable specimens. If the herbarium confirms a new find, a specialist from the state Department of Primary Industry conducts a risk assessment and, with stakeholders, develops a management strategy, which may include eradication—as in the successful recent elimination of an incipient water hyacinth invasion.

The European Environment Agency in 2010 published a technical report outlining a possible EDRR system for Europe. The report notes that the European Commission recognizes the urgent need for the commission and the individual nations to develop an appropriate information system for early detection and rapid response. The report calls for a response mechanism including rapid eradication (after risk assessment), continuous monitoring, and containment of potential spread. None of this has yet happened, undoubtedly because of the difficulties of effective European coordination on invasions (Chapter 8.2). Even without an organized early detection system, some invasions are found early in Europe, but often neither the individual nation nor the European Union acts to eradicate them. The spread of the small Indian mongoose (Chapter 8.2) epitomizes the problem.

The weakest link now in EDDR systems is simply the lack of enough trained personnel to seek out invasions. The trained weed spotters of Victoria give an indication of how citizen volunteers can be extremely useful in filling this gap. Even less formal systems can be very helpful, so long as there is sufficient publicity to sensitize the public to the problems associated with invasions, and in particular cases to show them what to look for. In France, for instance, the government engaged the diving and boating public in looking for the killer alga (Chapter 4.4), and private citizens were first to report several of the invasion events during the spread of that species. In California, the killer alga was recognized early because there had been so much publicity about the European invasion, and the early warning led to successful eradication. Posters about the killer alga have also alerted Californians to be on the lookout for other nonnative seaweeds. Eradication of the Asian longhorn beetle in Chicago region was possible only because an alert citizen collecting firewood recognized the insect from the widespread publicity surrounding the invasion of the same species in New York and notified authorities. In Vermont, a Boy Scout, whose troop helped remove invasive aquatic plants from Lake Champlain, spotted a patch of European water chestnut in a bog far north of its recorded range, and refuge managers extirpated it before the species spread to other bogs in the region. In Germany, a program recruits recreational SCUBA divers to identify and report aquatic invaders.

As is evident from the previous examples, even without massive organization, sufficient publicity about invasions and a well-publicized place to report them could be very useful in providing early warning. What is most needed is a phone number that any citizen could call or a website that anyone could use to report a possible invasion. A cell phone application could even be developed to send images to a central reporting entity, if such an entity existed. As might be expected from the description of New Zealand regulations in Chapter 8.2, this nation comes closest to effective early warning and does so partly through effective use of volunteers. There is a well-publicized Exotic Pest and Disease Hotline and trained technicians to respond to any report and decide whether further investigation is warranted.

9.2 When should an introduction simply be tolerated?

Only a minority of invaders cause substantial harm, and eradication or management of invaders can be very expensive. Further, some invasions recede on their own, like those of the Canadian waterweed in Great Britain or the giant African snail on Pacific islands (Chapter 4.4). These facts, plus a hesitancy to act in the face of uncertainty, particularly when action may be costly, often combine to encourage a wait-and-see attitude on the part of managers and policymakers confronted with a report of a newly recognized introduction. Under what circumstances is such a response justified?

Such spontaneous collapses are uncommon (Chapter 4.4), far less common than the opposite phenomenon, high-impact invasions after a time lag during which an introduced species is restricted and harmless. And even collapsing populations can leave a persistent damaging legacy, such as allelopathic chemicals. Clearly, there may be economic and environmental costs associated with both action and inaction. The question of what to do in any particular case would seem to be well suited to a risk assessment, if it were really possible to generate accurate predictions of what the risks are and what the probabilities are of their materializing. Many invasion impacts seem so idiosyncratic and surprising as to cast doubt on the ability to predict their possibility (Chapter 4.1), and, even if possible risks could all be identified, sound estimates of their probabilities are unlikely (Chapter 7.2). In addition, even if all the risks and their probabilities were well estimated, one would wish to know the likely cost of each (Chapter 7.2) to assess the likely net benefit of acting or not acting. But tallying the likely costs would be yet another difficult task involving guesses and approximation. Even aside from the probabilistic nature of the estimate, assigning "costs" in nonmarket situations renders cost–benefit analyses suspect. For instance, if a potential risk is extinction of a native species with no economic market value, how are we to determine the cost of such an event? Economists espouse various methods for estimating such costs, but all of them are indirect and controversial; a full discussion is beyond the scope of this book.

In many cases, a potential invader was discovered soon after establishment; however, nothing was done, and eventually a massive invasion ensued. The killer alga (Chapter 4.4) was found in a small patch just offshore of the Oceanographic Museum of Monaco before it had begun to spread. It could easily have been removed, but instead authorities argued. First they quarreled over who was responsible for its presence there, then over who should be responsible for managing it, and finally over whether it posed a sufficiently large threat to warrant management. By the time the last question was answered (in the affirmative), the alga had spread to thousands of acres off the coasts of several nations and eradication was impossible.

In another case, Koster's curse, a tropical American shrub, was found in a small area on Oahu in 1941, and it was restricted to this area into the 1950s. It was quickly recognized that this plant posed a threat, partly because of its invasive status elsewhere, but still authorities did not act, confident in their abilities to control it if it started to spread because a biological control project against this plant had succeeded in Fiji. In the mid-1950s the plant began to spread, the biological control insect was released, and it failed to hinder the spread. Koster's curse now infests 250,000 acres in Hawaii and is sometimes viewed as the state's second worst weed. It could almost certainly have been eradicated during the decade after its discovery. Similarly, in the continental United States, common crupina from southern Europe was first detected in Idaho in 1968, restricted to an area of 40 acres. Authorities did not act, pondering whether it would really become an invasive weed. By 1981 it had done so, infesting 22,300 acres. Only then did an eradication feasibility study begin, and it was not until 1991, by which time common crupina had been listed under the federal Noxious Weed Act, that a federal–state task force convened to discuss the problem. By then the plant had spread to four states and infested 62,000 acres; the project was abandoned. It probably could have been eradicated by manual means and herbicides when it was first discovered.

Cases such as these strongly suggest that, if it seems possible to eradicate a recently detected introduced species with little expense, it should be done. The precautionary principle, widely viewed as applicable to threats to the environment, states that if an action or policy risks causing harm and foreclosing options for future generations, in the absence of scientific consensus on the whether that risk is real, the burden of proof that it is not harmful falls on those taking the action. In this instance, the action is really the act of doing nothing, of not attempting the eradication. Invasions are often characterized as a situation with low probabilities of harm, but, if harm does ensue, the damage will sometimes be massive, as in the cases of Koster's curse, common crupina, and the killer alga.

What about situations in which an invasion has already happened and a species is already fairly widespread and either causing or threatening harm? If management seems warranted, what should be done? Broadly speaking, the options are twofold: (1) attempted eradication, just as for recently detected introductions; or (2) what is known as *maintenance management*, attempting to reduce populations of the invader to a sufficiently low level to minimize their impact and maintaining them at that level.

9.3 What is eradication, and when should it be attempted?

Eradication is the complete removal of every last individual from a distinct population, so that any recolonization of a site would have to come from another, spatially isolated population (if there is one). Politicians and popular writers frequently use the term *eradicate* incorrectly, calling for eradication of a pest when they simply mean elimination of most individuals. True eradication of invaders was long seen as a nearly impossible goal, and several high-profile early failed eradications seemed to confirm this view. The prevailing view through the 1990s was that, for most invasive species, eradication would be costly, probably futile, and likely to inflict collateral damage on nontarget organisms. Some famous failed eradication campaigns exemplify these problems. Most notorious was the 14-year effort, ending in the late 1970s, to eradicate the South American red imported fire ant in the southeastern United States. This project was such a legendary fiasco in terms of both collateral damage (including to humans) and expense (over $200 million) that publicity about it helped foster the environmental movement. This campaign was featured in Rachel Carson's seminal book *Silent Spring* and prominently denounced during the first Earth Day (1970). At the end of the eradication attempt, the fire ant had spread far more widely and was more numerous than when the project began. Even worse, the chemicals used to attempt to control the ant had many nontarget impacts (including potential effects on human health) and probably worsened the fire ant invasion by causing greater mortality for its natural enemies than for the fire ant itself. Subsequent research has shown that, from an ecological

Figure 9.1 Men trying to remove gypsy moths in a futile attempt to save the Dexter elm tree, Malden, Massachusetts, in 1895. (Photograph: WEBSITE: www.fs.fed.us/ne/morgantown/4557/otis/index_d.html.)

standpoint, the red imported fire ant invasion is limited by the fact that the species is a specialist of disturbed habitats, such as agricultural fields, and does not readily invade intact natural habitats, such as forests.

Another high-profile failed eradication campaign was initiated in 1890 by the state of Massachusetts against the gypsy moth. The introduction and spread of the moth were described in Chapter 6.6 and its impacts in Chapter 3.4. For 10 years, the state employed thousands of men in a scorched-earth approach using chemicals, fire, felling of massive numbers of trees, and hand collection of countless egg masses, caterpillars, and pupae. The campaign was finally abandoned when it became apparent that, in spite of tremendous expenditures, the moth was spreading rather than receding.

A long, disastrous eradication campaign targeting white pine blister rust in North America (Chapter 5.3) began in the early 20th century. Because ground cover plants in the currant and gooseberry genus serve as alternate hosts of this fungus, the effort mainly focused on an attempt to eradicate these plants, including native species. These plants are very common and widespread—a fact that should have generated doubt about the feasibility of the project from the start. However, the commercial value of timber from affected species in the white pine group generated enormous pressure to attempt the impossible. The largest expense was labor (a total of over $150 million), especially during World War I, although after the war currant and gooseberry eradication served as an excellent source of employment for returning soldiers. During World War II, German and Italian prisoners of war and inmates of several American prisons were even engaged in the grueling work, which primarily involved hand digging and distributing chemicals. At first, the alternative host plants were eliminated within 100 yards of pines, then within 200 yards, and then 600 yards, and finally they were targeted for total elimination everywhere, including in wetland and riverine habitats. The first focus was on introduced European black currant. The density of this species was reduced, and the reduction probably temporarily reduced the transmission of white pine blister rust to pines. However, subsequent efforts to eradicate all the native currants and gooseberries were ineffective and had little if any impact on spread of white pine blister rust to pines. These efforts also caused massive collateral ecological damage, such as removing all vegetation from streamsides with a special rake, then reseeding with grass.

Even in the face of apparent failure, the white pine blister rust eradication was not terminated for decades, but its goal gradually shifted to simply lowering disease rates for pines rather than total eradication of the rust. Why did an ineffective, expensive program drag on for so long after it was apparent that it was not working? Probably there were three reasons, versions of which were shared by the failed fire ant and gypsy moth eradications. First, the fungus was killing many trees, so the economic impact was enormous; ecological impacts in general did not cause much public concern in the first half of the 20th century, but money was a different matter. Second, in the face of adversity, most people are motivated to try to do something, almost anything. Americans in particular have long prided themselves on having a can-do spirit. Historically, many big projects have appealed to this spirit, especially in times of hardship, such as during the Depression and World War I and II. Finally, when any project gets as big as the white pine blister rust eradication campaign, it develops its own bureaucracy and momentum, with stakeholders whose interest is to maintain the project for their own sake, whether or not the original goal is served. Inertia therefore becomes enormous.

The failures have often been publicized, but there have also been many successes—over a thousand of them worldwide—and these tend not be very well-known. Perhaps the earliest insect eradication was the elimination of tsetse flies from the island of Principe in the Gulf of Guinea. The flies were introduced in cargo from Africa in 1825, bringing with them sleeping sickness that appeared 35 years later, ultimately reducing the human population tenfold. A four-person team managed to eradicate the fly (and the disease) completely between 1911 and 1914. In 1956, tsetse flies were again detected, and a large scientific team was immediately dispatched to the island, where they captured 66,894 flies in two months. Using traps, insecticides, brush clearing, and hunting to reduce populations of feral pigs and dogs (alternate targets for the flies and hosts for the sleeping sickness pathogen), they eradicated the fly once again.

Introduced insects have similarly been eradicated from various other islands, always when they threaten agriculture or public health. The New world screwworm fly was eradicated from Curaçao in the Lesser Antilles by the remarkable sterile male technique, pioneered by the entomologist Edward F. Knipling (1909–2000) of the United States Department of Agriculture (USDA). Knipling conceived the notion of mass-producing large populations of a target pest, separating the males, sterilizing them by irradiation but not killing them, then releasing them in the midst of the target population. The theory was that, by chance alone, if females mate with males randomly, they will be highly likely to mate with sterile males and thus produce no offspring, so long as they mate only once per season and enough sterile males can be produced that they vastly outnumber fertile ones. Further theory suggests that, if the population can be driven below a certain threshold by this means, it will ultimately decline to extinction.

The test case for Knipling's brainchild, eradication of the screwworm from Curaçao, worked exactly as he had hypothesized, and the technique has since been applied elsewhere. For instance, the melon fly was eradicated from Rota Island, near Guam, by this technique. The sterile male technique was subsequently used to eradicate the screwworm from the United States, Mexico, and part of Central America, and it has been used in many locations to eradicate outbreaks of other insect pests. In some projects where eradication has not been feasible, such as the attempt to control the tsetse fly in Africa, the sterile male technique has nevertheless been used repeatedly to control pests at low levels. With various modifications, the technique has been attempted in projects targeting species other than insects, such as the sea lamprey in North America. Mass rearing in sufficient numbers and sterilization that is sufficiently foolproof are difficult technological challenges to sterile-male efforts, but they have often been surmounted. For instance, radiation and chemicals can be used for sterilization. An occasional problem is that sterilization produces, as a side effect, males that are less likely to mate with females (because either their own efforts are insufficient or females somehow detect that they are different), countering the advantage of great numbers. This problem has plagued some attempts to control the Mediterranean fruit fly (medfly). In a campaign to eradicate the melon fly from the Ryukyu Archipelago, sterile males were both less competitive and less attractive to females than wild males, but not to the point of preventing the project from succeeding.

A related approach, known as the male-annihilation method, entails killing a large enough fraction of the males in a population that the number of mating females is insufficient to sustain the population. This approach rid both Guam and Rota of the Oriental fruit fly. More recently, the melon fly was eliminated from the entire Ryukyu Archipelago, including the large islands of Okinawa and Amami, by a combination of the sterile-male and male-annihilation techniques.

On small enough islands, even brute-force methods may succeed if applied consistently and assiduously. On Key West in the 1930s, the Asian citrus blackfly was eradicated simply by three years of spraying with a mixture of paraffin oil, whale oil, soap, and water. The cactus moth from South America was eradicated from Isla Mujeres in Mexico by complete removal of all the cactus host plants. Islands present huge advantages in terms of eradication attempts for two reasons. First, if they are small, the population is small and the effort will be less expensive. Second, if they are isolated enough, continued immigration (or reinvasion, in case of a successful eradication) is less likely than on the mainland. A mainland campaign, identical to the one that succeeded against the Asian citrus blackfly on Key West, failed in Florida largely because these advantages were absent.

Nevertheless, introduced insects have been eradicated from large islands and even the mainland. As mentioned already, the large islands of Okinawa and Amami were rid of the melon fly. Even more remarkably, an African mosquito that vectors malaria was eradicated from 21,000 square miles of northeastern Brazil in 1939–1940 through chemicals deployed against both adults and larvae. Insects that have been discovered soon after introduction have often been eradicated from small areas on large islands or continents. For instance, the white-spotted tussock moth, an Asian native related to the gypsy moth, established a population in the vicinity of Auckland, New Zealand in 1996, and the New Zealand government acted quickly and successfully to eradicate it, using aerial and ground spraying of the bacterium *Bacillus thuringiensis* (Bt) the first year. During this time, a team of New Zealand and Canadian scientists identified a chemical (a pheromone) that female moths produce to attract males. They learned how to synthesize it, and 6,500 pheromone-baited traps were employed during the second year, killing the last moths.

Eradication campaigns against two recent high-profile insect invaders in the United States (Chapter 6.6) have had various results. In the Chicago area, the attempt to eradicate one of the incipient invasions of the Asian longhorn beetle succeeded after nine years and in New Jersey eradication took 11 years, while eradication attempts are still under way against incipient invasions by the same insect in New York and Massachusetts. The strategy in all these areas has been to quarantine wood from infested areas that houses the beetle and to locate, fell, chip, and burn infested trees. The equally infamous emerald ash borer was the target of poorly coordinated and inconsistently implemented eradication campaigns in Ohio and Michigan, both of which failed. The emerald ash borer has now spread as far east as Maryland and as far south as Kentucky.

As we see, many eradication campaigns against insects have succeeded, but many have failed. A discussion of what separates the failures from the successes will be found in the next section, after consideration of eradication efforts against other species. The giant African snail, whose introduction and impact on Pacific islands were discussed in Chapter 6.1, was brought in 1966 to Miami from Hawaii by a young tourist who presented several to his grandmother as a gift. She released them in her backyard, and an invasion covering 42 city blocks was discovered in 1969. The eradication campaign used a quarantine, a chemical bait, and hand-picking. Despite the finding of several satellite invasions, the eradication succeeded in 1975, and this success inspired the eradication of a similar giant African snail invasion in Queensland, Australia. In Miami, a new invasion was detected in 2011, triggering a new eradication campaign, and over a thousand snails were quickly removed from an area of 1 square mile.

The Australian government conducted another successful eradication campaign, this time against the Caribbean black-striped mussel, a relative of the zebra mussel. This mollusk was discovered in 1999 in Cullen Bay (31 acres), a part of Darwin Harbor, within six months of its arrival and before it had spread further. Within days after the discovery, the bay had been quarantined and treated with 42,000 gallons of liquid bleach and many tons of copper sulfate, a molluscicide. All mussels, and individuals of many other species, were killed, though the latter species have all recolonized the area.

Many invasive mammal populations have been eradicated, especially on islands. A nongovernmental organization, Island Conservation, has as its raison d'être the eradication of invasive mammals that threaten native biodiversity on islands. Island Conservation has so far contributed to 66 successful eradication operations on 48 islands. The governments of New Zealand, the United States, France, and Australia have also been particularly active in mammal eradication projects. Among main targets, 318 populations of ship rats, Norway rats, and Pacific rats have been eradicated on islands around the world, out of 344 attempts. In most instances, the goal of rat eradication has been to save populations of seabirds nesting or roosting on islands, and the size of islands from which rats have been eradicated has increased greatly as technologies have improved. The largest success to date has been the eradication of Norway rats from subantarctic Campbell Island, 44 square miles, but far larger rat eradication projects are in the planning stages. Toxic baits are the key technique, and the experience of many projects has led to several great improvements, particularly in the efficiency of deploying and situating baits and in using baits and traps that minimize nontarget impacts (such as secondary poisoning of animals scavenging dead rats). In several instances, rat eradication has triggered explosions of house mouse populations that had been suppressed by rat predation, with negative effects of the mice similar to those that had been caused by the rats (e.g., predation of seabird eggs). This experience has inspired scientists to develop techniques that eradicate rat and mouse populations simultaneously.

Great Britain experienced an early triumph in its 10-year campaign to eradicate 200,000 nutria from Great Britain; success was declared in 1981. The main technique was trapping, and the key to the success was careful study of the population biology of the nutria and meticulous mapping of populations and trap results. This project led to an ambitious campaign, begun in 2000, to eradicate several hundred thousand nutria from the Chesapeake Bay area, where their feeding on marsh plant roots has led to erosion and the loss or marshlands to open water. The outcome of this campaign, engaging federal and state agencies and private organizations, is still in doubt. In the Galapagos Archipelago, pigs have been eradicated from Santiago Island (143,000 acres) through a combination of poisoning and hunting on the ground. Goats have also been eradicated from Santiago Island, through ground hunting with dogs, followed by aerial hunting and a modification of the traditional Judas-animal technique. The latter is a widely used approach in eradicating or managing animals that tend to band together. An individual is trapped, fitted with a tracking device (originally a bell, now a transmitter), and released; hunters follow it to find the other individuals. For goats, Karl Campbell of Island Conservation developed super-Judas goats by sterilizing females and injecting them with hormones that keep them in a long-term estrus. This method allowed hunters to track and kill the last of over 89,000 goats removed from the island. The same approach is now being followed in a campaign to eradicate goats from 1,131,000-acre Isabela

Island. Other introduced mammals that have been eradicated from islands include the muskrat in Scotland by a government campaign and various combinations of mice, European rabbits, sheep, burros, and feral cats on islands in the Gulf of California by Island Conservation.

In general, eradication of aquatic animals has been more difficult than that of terrestrial ones. However, several small populations of introduced fish have been eradicated, such as the European gudgeon from the Auckland, New Zealand, region and the South American pirambeba and Central American three-spot cichlid from Florida. Small populations of the Australian hairy marron crayfish have been eradicated from the vicinity of Auckland, New Zealand. However, once an aquatic species moves beyond an enclosed location like a lake or a lagoon that can be drained or chemically treated, the challenge is far greater and successes are rare.

Plants also have a reputation as unpromising targets for eradication, largely because many of them have long-lived soil seed banks. However, many small populations of plants have been eradicated, and even a few larger ones. Among terrestrial plants, sandbur was eradicated from the small Hawaiian island of Laysan by use of hand weeding and chemicals for several years to exhaust the soil seed bank. More recently, the Old World pasture weed kochia was eradicated from thousands of acres spread over 560 miles in Western Australia. In the southeastern United States, the African parasitic plant witchweed (Chapter 3.5) has been the target of a 50-year eradication campaign that has reduced the infested area from over 400,000 acres to 6,200 acres, and there is every reason to believe it will eventually succeed. On a much smaller scale, the California Department of Food and Agriculture has eradicated over 1,000 isolated populations of 14 introduced invasive weeds, including serrate spurge, giant dodder, and Austrian peaweed.

Several small introduced populations of aquatic plants have been eradicated. Perhaps most noteworthy is the eradication in southern California of two small invasions of the killer alga (Chapter 4.3). Each was much larger than the original European invasion described already, but in contrast to European authorities California agencies and private industry acted quickly, forbidding boat traffic to stem the spread of the invader and using chlorine pumped under tarpaulins that covered patches of algae. Interestingly, although the European failure received massive international publicity, the successes in California were reported only locally. Also in California, in a 16-year project hydrilla was eradicated from a 158-acre lake through the use of quarantine, chemicals, divers with suction dredges, and a drawdown.

The year 2011 saw one of the great triumphs of eradication: rinderpest, the viral disease of cattle that had ravaged native African mammals with many ecosystem consequences (Chapter 3.5), was declared eliminated globally. The United Nations held a ceremony to mark the accomplishment, the culmination of a campaign that began in earnest in 1994. The main technique was engaging pastoralists in vaccinating livestock over vast stretches of Africa.

Eradication of invaders frequently leads to surprises. As was mentioned already, eradication of rats has occasionally led to explosive growth of previously suppressed house mouse populations. Rat removal has also occasionally released introduced herbivore populations that go on to cause great damage. For example, eradication of Pacific rats from Motuopao Island (New Zealand) to protect a threatened native snail led to greatly increased populations of the introduced common garden snail, to the detriment of the native species. Eradication of introduced herbivores has sometimes led to remarkable growth of previously innocuous weeds. On Motunau Island (New Zealand), rabbit eradication led to proliferation of introduced boxthorn, and eradication of large grazers from California's Santa Cruz Island resulted in explosive growth of Eurasian fennel and other introduced weeds. These have gradually receded as native vegetation has recovered. On Sarigan Island in the Northern Marianas, eradication of introduced goats and pigs led to a totally unexpected massive growth of a South American vine, vine paper rose. Again, as native vegetation has recovered, infestation by the invader has declined.

Perhaps the most shocking surprise was the crash in the California Channel Islands of the endemic Channel Islands fox, caused by the eradication of large populations of introduced pigs on two of the main islands. Golden eagles, which had been attracted to the islands originally by the pig populations, then switched to eating foxes, threatening their very existence. In another surprising turn of events, on subantarctic Macquarie Island, eradication of feral cats that had threatened bird populations led to an unexpected explosion of introduced European rabbits. The rabbits stripped much of the island of vegetation, to the great detriment of the very birds the cat eradication was supposed to protect (Chapter 4.1). A similar surprise resulted from the eradication of Norway rats on Canna Island, in the Scottish Hebrides—an explosion of introduced rabbits ensued. The take-home message from these surprises is to attempt to predict such events, primarily by careful consideration of the role of the targeted invader in the food web.

Finally, an ongoing development in New Zealand is the creation of mainland islands. Patches of good habitat on the large main islands are enclosed with an invasion-proof fence, after which invasive mammals within the fence are eradicated, and the fenced, invasion-free area is gradually expanded. Ultimately, native species threatened by the eradicated invaders are reintroduced. The largest such project to date is Maungatautari, which now encloses 8,400 acres, including a large mountain, and has 30 miles of fence. A nongovernmental organization has eradicated 11 of 14 invasive mammals originally at this forest site: red deer, brushtail possums, European hedgehogs, ferrets, stoats, weasels, goats, boar, cats, ship rats, and Norway rats. European hares and rabbits remain in small areas and are likely to be eradicated soon, but house mice have been a persistent problem, particularly because they can be removed only by a bait-delivery system that does not deliver poison to nontarget species. Several native bird species that had been eliminated by the introduced predators have been reintroduced and thrive, and native plants that had been suppressed are now flourishing.

9.4 What characteristics separate successful eradications from failures?

Are any features common to successful campaigns? Do certain kinds of problems often cause failures? Six aspects of each case seem critical to the chances for success.

9.4.1 Plan for the entire project

It is important that resources be committed at the outset to see an eradication project through to completion. This issue becomes particularly important in the final stages of an eradication attempt, as it often costs more to eliminate the last 1% of an invasive population than it does to eliminate the first 99%. Furthermore, as an eradication program proceeds, the target population falls, and the perceived urgency of dealing with the invader may lessen in the eyes of funding agencies or the public. Often small, restricted populations can be eradicated with little cost, but for widely distributed targets, an eradication, even if feasible, can be extremely expensive. The previously mentioned screwworm eradication in the United States and Mexico cost about $750 million, and the ongoing campaign to eradicate witchweed in the South has cost about $250 million. Compared with the costs of not eradicating, such large costs may well be justified in certain cases. For instance, every year the United States government spends $5 million trying to control the brown tree snake on Guam and keep it from spreading. Would it not make sense to explore the possibility of spending, say, $100 million in a campaign to eradicate it for once and for all? However, agencies and policymakers must be convinced of the worth of the entire project and recognize the likely expense before the project commences.

9.4.2 Lines of authority

Eradication is an all-or-nothing process; it cannot succeed if some individuals are able to prevent whatever operation is deemed necessary. Thus, for example, a program to eradicate a weed species will founder if a landowner whose property contains the weed refuses to permit the eradication on the property or refuses to allow the use of herbicide if that is the method being used. In certain situations, some stakeholders prize the target species and do not want to see it removed. For instance, hunters have objected to boar eradication schemes in Hawaii, and eradication of invasive eucalyptus trees from Angel Island, California, was opposed on the grounds that the trees were beautiful, provided shade, and had long been present. Other times, killing vertebrates by hunting, trapping, or poisoning raises opposition from animal rights groups. The attempted eradication of the North American eastern gray squirrel in Italy was halted outright by the courts acting on complaints by animal rights advocates, and other eradication campaigns for mammals, birds, and even trees have faced similar objections (Chapter 11.6). Or individuals may object to possible nontarget impacts of an eradication method. For instance, aerial spraying of malathion to eradicate medfly invasions in California and Florida generates widespread complaints of discomfort or danger to health. Attempts to stem citrus canker in Florida by cutting down citrus trees led to vociferous objections from homeowners who prized their apparently healthy ornamental fruit trees.

Some islands that have seen successful eradications are uninhabited by humans or have small populations; many are remote from popular news sources. These features have probably facilitated eradication. As eradication projects move to ever-larger stages, some entity—a government agency, in most cases—will have to able to compel cooperation. Undoubtedly, such cooperation will be aided by a thoughtful and substantial program to educate the public about the damages caused by the invader, the benefits that would flow from its eradication, and the appropriateness of the technique chosen. For instance, some animal rights groups are willing to countenance sterilization campaigns to eliminate invasive vertebrate populations, but the public must be made to realize that no such technology is currently operational.

9.4.3 Biology of the target species

Some traits of invasive species make eradication feasible or much less likely. Certain such features are obvious: it is much easier to find all the individuals in an invasive population of some large mammal (say, deer) than in an invasive population of a tiny insect (say, an aphid). Other times, only substantial research can elucidate weak points in the life cycle of a target species. For example, eradication of the giant African snail was greatly aided by the fact that it does not self-fertilize, and the eradication of the malaria mosquito in Brazil was possible largely because it was found almost exclusively near buildings. Perhaps the most famous eradication of all, the global elimination of smallpox in the 1970s, was feasible only because smallpox has no nonhuman reservoir or long-term carriers.

9.4.4 Probability of reinvasion

One reason so many eradications have been attempted on islands is that, in the case of success, reinvasion is unlikely. By contrast, if a population of some mainland invader is eradicated, the same site may be recolonized quickly by individuals from contiguous or nearby populations. For example, Eurasian watermilfoil was eradicated from a large lake in Washington State; however, the presence of a public boat ramp led to rapid reintroduction from boat propellers, and the operation was shifted to one of maintenance management by manual removal. Subversion of an eradication by opponents or sociopaths is also a possibility; it is suspected in the failure of a project to eradicate northern pike from a lake in California (Chapter 6.1). It is possible that, in certain circumstances, economic benefits could justify eradication even if fairly rapid reinvasion were certain. For instance, expensive eradication campaigns are repeatedly launched against the medfly in California and an Asian variety of the gypsy moth in both the United States and Canada, despite continued reinvasion. In both cases, if the insect were to establish a population, trade regulations (for fruits and vegetables in the case of the medfly and for timber in the case of the gypsy moth) would either forbid export of affected species or require expensive treatment. The economic cost in both instances would be so great that it pays federal and state authorities not only to be extremely vigilant but also to eradicate incipient invasions as soon as they are detected, even though the absence of the pest will surely be only temporary.

9.4.5 Who bears the costs and who reaps the benefits?

Such cases would seem to be ideally suited to cost–benefit analyses, but these analyses depend on who bears costs and who estimates them (Chapter 7.2). For instance, for the medfly invasion, taxpayers pay for the eradication campaigns, which primarily benefit growers. Furthermore, as previously mentioned, aerial spraying of malathion to eradicate the medfly is claimed to have human health impacts and certainly induces temporary discomfort. In 2008 and 2009, a controversy erupted in California over a federal and state agency project to eradicate the light brown apple moth, a potential crop pest from Australia. The aerial spraying of an attractive pheromone (Chapter 10.5) was claimed to have affected human health, and organic farmers in sprayed regions incurred major losses because their produce could no longer be classified as organic.

9.4.6 Possibility of restoration

Eradication, even if successful, does not by itself constitute ecological restoration. Ecosystems are constantly changing in response to environmental changes and evolution of component species, but various exogenous forces, such as introduced species or human modification of the landscape, can radically disrupt the natural trajectory of ecosystem change. Ecological restoration consists of returning a damaged ecosystem to its historical trajectory. In campaigns to eradicate agricultural pests or threats to human health, restoration is not a goal, but eradication for conservation or ecological purposes is usually either explicitly or implicitly associated with a desire to put a damaged ecosystem on the road to recovery. The precise meanings of *natural state, recovery*, and even *restoration* are technically challenging, and a full discussion is beyond the scope of this book (however, see Chapter 11.5), but most programs aimed at eradicating or managing invaders for conservation envision the eradication of nonnatives as serving the goal of restoration in a commonsensical interpretation of the term. Strictly speaking, this may sometimes be impossible. For instance, the invader may have driven to extinction a native species, so no recovery of that population is possible. Or eradication may succeed, but the native species expected to recover may nevertheless not do so, for unknown reasons. For example, after successful eradication of predatory mammals on several New Zealand islands, the reintroduction of the stitchbird long proved difficult, although several island populations now exist. Also, as noted already, eradications are sometimes followed by unintended consequences (for instance, explosion of mouse populations after rat eradication or proliferation of an introduced weed after an introduced herbivore is removed), and these may impede restoration. The key point is the necessity to plan the eradication in the context of a restoration project, to anticipate difficulties, and to allow for contingencies such as unintended consequences. In some cases restoration will happen by itself (for example, the return of massive seabird colonies after island rat eradication), but often additional measures are needed.

10

MAINTENANCE MANAGEMENT
OF INVASIONS

If eradication fails or is not attempted, it may still be possible to maintain invasive species at low levels and minimize their impacts. Many invasions have been successfully managed in the long-term by a variety of means. These tactics mostly fall in three categories, mechanical or physical control, chemical control, and biological control, though it is not uncommon for the first two to be combined. In addition, ecologists have suggested managing entire ecosystems in such a way as to inhibit introduced species generally and to favor native species rather than focusing on managing invasive species one by one, as in traditional maintenance management. An approach used in agriculture on a number of invasive insects, termed *integrated pest management*, attempts to combine chemical control with aspects of ecosystem management and biological control.

10.1 What are mechanical and physical control?

Physical control can be as simple as hand-pulling weeds or as complicated as using equipment such as modified backhoes and bulldozers for the same purpose. Between these two extremes are such hand tools as root wrenches and

Pulaskis and portable mechanized tools like weed whips and chainsaws. Use of powered tools or machines is often termed mechanical control, while physical control consists of using one's hands and simple unpowered tools. Many highly damaging invasive plants have been successfully managed by mechanical or physical means. In the state of Kentucky, supervised teams of volunteers convicted of driving under the influence of alcohol keep Eurasian musk thistle under control in state parks and preserves. European beach grass was long maintained at minimal levels at the Humboldt Bay National Wildlife Refuge without use of herbicide through physical removal by refuge personnel, employees of a state public works program, and prison work crews. The initial phase was difficult because a huge amount of beach grass had accumulated (Figure 10.1 top), but subsequent maintenance requires far less labor (Figure 10.1 bottom). In many places in the American South, moderate infestations of kudzu have been adequately controlled by mechanical or physical removal, sometimes combined with prescribed fire. Again, the beginning of such programs is often difficult because of the large amount of accumulated vines, but subsequent maintenance mostly consists of monitoring for new growth and a modest amount of spot removal.

Another approach consists of covering invasive plants with sheets of landscaping material to raise the soil temperature to kill invaders (termed *soil solarization*). This method was first developed as a control method in agriculture for fungi, bacteria, nematodes, and soil arthropods, both native and introduced. It is sometimes an effective control method for invasive plants as well. The covering can remain from weeks to two years, depending on the target plant and the environment. In Northern Ireland, cutting, then solarizing common cordgrass was as effective as treating it with herbicides. A general downside of this approach is that, especially for large areas, maintaining the covering is expensive and sometimes technically difficult.

Figure 10.1 (Top) Dune area in Lanphere Dunes Unit, Humboldt Bay National Wildlife Refuge Complex, California, 1992, overrun by European beach grass. (Bottom) Same site, 2012, after beach grass removal between 1992 and 1996. (Photograph courtesy Andrea Pickart.)

Some invasive animals have also been managed by physical or mechanical means. Recreational hunting or commercial hunting and trapping are sometimes employed as one tool to keep populations of invasive mammals from increasing—boar and boar–pig hybrids, goats, and donkeys have been among the targets. More frequently, hunting and trapping have been combined with chemical baits, and some such programs are remarkably effective. A striking example is the Alberta Rat Patrol, a small team of specialists who enforce provincial regulations requiring property owners to rid their properties of rats (or have the Rat Patrol step in, at a hefty fee), distribute anticoagulant baits, and aggressively hunt Norway rats in such sites as garbage dumps. The program is so successful that Alberta touts itself as rat-free, and this is not far from the truth; rat sightings are very rare.

Many successful programs to maintain populations of invaders at low densities rely on heavy use of labor, which is often a strike against them. Two sources of free labor are increasingly employed. Prison work crews are used to remove invasive plants, as was noted for musk thistle and beach grass. A political and philosophical issue is whether such crews should be used to compete with paid workers. This is a valid concern, but it is difficult nowadays to find projects that have used prison labor that would simply have hired workers if prison work crews had been unavailable. It is more likely that the projects would not have been undertaken in the first place. A serious practical matter is that such work crews need strict supervision, and not only to prevent prisoners from escaping. Sometimes invasive plants are not easily distinguished from desirable native species, and a scorched-earth approach of ripping out all vegetation would be counterproductive. Training and supervision may be sufficiently expensive to outweigh the benefits of a massive amount of prison labor.

In wealthy nations, volunteer labor may help minimize populations of invasive plants. In the United States, the Nature Conservancy has organized many such campaigns, while the

Grand Canyon Wildlands Council, a nongovernmental conservation organization, has mounted major programs to reduce salt cedar invasion in the Colorado River and its tributaries. In Florida, the Pepper Busters, a volunteer group, has aided the state's effort to contain Brazilian pepper, Florida's worst invader. There is certainly a large pool of volunteer labor available for various environmental projects, and the growing visibility of the damage caused by invasions makes it easier to engage citizens in this activity. In the United States, ecotourists pay substantial sums to spend part of their vacation pulling up invasive plants in scenic locations like national parks. As with prison work crews, a major factor in this approach is the necessity of training and coordinating teams of volunteers to avoid nontarget impacts, but, unlike supervision of prison crews, escape is not an issue. Volunteer efforts may also serve another purpose—a motivational one. By engaging many citizens directly in the effort to control invasive species, authorities may raise awareness of the issue in general and educate the public about ways to minimize damaging invasions, all the while tapping a useful source of labor. For instance, in Victoria, British Columbia, broom bashes sponsored by the Garry Oak Meadow Invasive Plant Removal Project to remove Scotch broom are so popular that they are listed in a local monthly environmental newsletter, and the project attracts many young people, such as schoolchildren and Girl Guides.

In addition to volunteers and prison work crews, paid labor can be used in managing invasive species. The largest, most sustained, and most comprehensive such program is South Africa's Working for Water program, a massive public works effort initiated in 1995. This project has provided jobs and training to about 20,000 people per year (more than half of them women) who would otherwise be underemployed or unemployed and has cleared more than 2 million acres of invasive introduced plants such as several species of mesquite and acacia. Much of this work is simple but grueling physical control—cutting down and ripping out invasive plants. Chemical and biological control also play roles.

The combination of physical or mechanical control and chemical control is often used to manage plants, just as it is for some mammal eradication and maintenance management programs. In South Africa, for instance, a combination of mechanical and chemical control is often successful against tropical American lantana. A major campaign in south Florida against Australian paperbark, employing primarily mechanical and chemical means, but with some biological control, has reduced areal coverage by over half. This coordinated state program uses aerial spraying of herbicide over vast expanses of paperbark-dominated forest as well as manual felling of trees in many areas followed by application of herbicide to cut stumps. In this campaign, insects have also been released for biological control, with the hope that they will bring about control on infested private lands, where state law prevents state resources from being expended.

10.2 What are the advantages and disadvantages of chemical control?

The use of herbicides, rodenticides, piscicides, molluscicides, and insecticides, including in baits, is sometimes remarkably effective in retarding invasion, aiding eradication, or providing long-term maintenance management. Chemical control is often also highly controversial, even when effective, because of possible nontarget impacts (including on human health) and expense. The publication in 1962 of Rachel Carson's *Silent Spring*, which helped found the environmental movement, was largely an attack on the use of pesticides and featured campaigns against two major invaders in the United States: the red imported fire ant and the gypsy moth. Carson, who was vilified by the chemical industry and the United States

Department of Agriculture (USDA), was largely vindicated by scientific evidence on the nontarget impacts of pesticides, particularly chlorinated hydrocarbons. Part of Carson's legacy is much more stringent testing and regulation of pesticides and herbicides, and another part is a suspicion among much of the public, sometimes bordering on chemophobia, about any chemical control method.

If used properly, many pesticides and herbicides now available are far less likely to have nontarget impacts than the broad-spectrum pesticides that Carson lambasted. They are not always used properly, but when they are they can sometimes be very effective alone or in combination with other methods. In some instances they are the only effective method currently available. For instance, fluridone, an herbicide that inhibits synthesis of pigments and is widely used against a variety of aquatic plants, is currently quite effective against small and medium-sized infestations of hydrilla, and at least against the latter no other approach has given good control.

Water hyacinth in Florida has been reduced from about 126,000 acres in 1960 down to a maintenance level of about 1,000–2,000 acres each year, primarily by use of the herbicide 2,4-D, a synthetic hormone that regulates growth. Mechanical harvesters had occasionally been used but were unable to keep up with the fast growth of water hyacinth. Two South American weevils that fed only on water hyacinth were released in the 1970s and established populations, but they failed to control the plant (although they have worked better in Africa, substantially controlling water hyacinth in Lake Victoria). However, 2,4-D was one of the two components of Agent Orange, a controversial herbicidal mixture sprayed in huge quantities over Vietnam during the Vietnam War to destroy forest vegetation and deprive guerillas of cover. Agent Orange turned out to have major impacts on the health of humans (i.e., causing birth defects and cancer), livestock, and probably wild animals, and its use is now widely seen as a grave error. The toxic effects have largely been linked to dioxin, a contaminant and by-product of the manufacture not of 2,4-D but of the other component of Agent Orange. However, the inevitable association in people's minds of 2,4-D with Agent Orange has led to tremendous wariness with respect to its use, reflected in stringent regulations and sometimes bans. The key feature in the Florida water hyacinth campaign is that, although initially large amounts of 2,4-D were used to reduce the massive infestations, within a few years these were gone, and the ongoing maintenance management requires less than 1% of the amount originally used.

Insecticides have a long and often lamentable history, parts of which were highlighted by Rachel Carson. For instance, attempts to eradicate or simply to control gypsy moth in the United States first used Paris green (copper acetoarsenite), which was toxic to plants, and then used trucks to spray lead arsenate, which was toxic to animals. In the mid-1940s, aerially sprayed DDT replaced lead arsenate. Concern about nontarget impacts of DDT (including human health impacts) was common in the 1950s and was only heightened when it was found in the milk of dairy cows. Carbaryl, another broad-spectrum insecticide (that is, one that can be used against many species), largely replaced DDT in the 1960s, and trichlorfon, yet another broad-spectrum insecticide, was added to the armamentarium in the 1970s. Both of these chemicals affect the nervous system of most insects but also some other animals, so these were replaced by diflubenzuron, which inhibits the growth of insect exoskeletons. Unfortunately, it affects other insects and arthropods generally, not just gypsy moths, and it is sufficiently persistent that it can affect the moth and nontarget species alike months after application.

Since the 1980s, the bacterium *Bacillus thuringiensis* (Bt) has been cultured commercially and used to attack local gypsy moth infestations. Although not as broad-spectrum as the chemicals it replaced, Bt does attack the gut of other moths and butterflies in addition to gypsy moth, so long as their caterpillars eat sprayed leaves. Bt spraying must also be carefully timed, as young gypsy moth caterpillars are much more sensitive to it than older ones. A viral disease that affects only gypsy moth caterpillars has occasionally been used, particularly where the potential nontarget impacts of Bt are unacceptable (e.g., if an endangered moth or butterfly species might be affected). However, the viral insecticide is too expensive and difficult to produce for wide-scale use. The upshot is that nontarget impacts are still a major issue in the use of insecticides against gypsy moth, a century after they were first deployed.

Insecticides have also been used for almost a century to battle the Japanese beetle in the United States, first with lead arsenate and lime and then with sodium cyanide, which is highly toxic to both invertebrates and vertebrates. In the 1950s DDT was aerially sprayed against the Japanese beetle, as was dieldrin, another broad-spectrum insecticide. Carbaryl and chlordane, also broad-spectrum insecticides, have also been used against the Japanese beetle and have succeeded in eliminating local infestations, and more recently a soil drench of imidacloprid was employed against the larvae; imidacloprid affects many other soil insects and has unknown long-term ecological consequences. Two bacterial diseases are commercially available for use against Japanese beetle larvae—milky spore and a strain of Bt that attacks Japanese beetles. They are expensive, and neither has been used in widespread management projects; however, both have been useful in control by homeowners of some small infestations.

Chemical insecticides are much more widely used in agriculture than for maintenance management of invaders that are environmental pests. Concerns about environmental impact are usually not paramount in agriculture, although nontarget impacts, especially on human health, can lead to regulation or even banning of insecticides that would otherwise provide adequate long-term control if routinely used. However, chemicals are sometimes employed in the service of conservation, at least as a stopgap measure in the hopes that, ultimately, some other approach (such as biological control) will be developed.

An excellent example is the management of the yellow crazy ant to save the endemic red land crab of Christmas Island (Chapter 4.2). The remote location and rugged topography of the island had hindered all attempts to manage this invasion, and the red crab seemed destined for extinction. However, the use of helicopters by Parks Australia beginning in 2000 to disperse tons of bait containing the chemical fipronil has greatly limited the ant and led to a partial recovery of the crab population. This is probably not a long-term solution for two reasons. First, it is very expensive, particularly because of the isolation of the island. Second, fipronil is a broad-spectrum pesticide affecting the insect nervous system and also that of crustaceans such as the red crab. Detrimental effects on some vertebrates are also suspected. Nevertheless, if the red crab, whose population was dwindling rapidly, were to be rescued, no other method was available, and the full impact of fipronil in this campaign, including possible long-term effects, is not yet known. However, it is unlikely that any conservationists would object to this rescue, even though the long-term goal is a less problematic and costly solution.

Chemicals have also been widely used to control invasive fish and mollusks. Chemical control of mollusks has largely been spurred by the attempt to control several diseases of humans and other vertebrates, such as schistosomiasis and fascioliasis, for which snails are intermediate hosts. Some chemicals used as insecticides also have been used against snails, such as carabaryl and malathion. Several synthetic chemicals have been developed specifically as molluscides, however, and an extract from the fruit of chinaberry (a south Asian native that is invasive in several other parts of the world) has been locally used as a molluscicide. Molluscicides have been used to attempt to manage several invasive nonnative aquatic mollusks, such as the golden apple snail and the zebra mussel, but the expense and nontarget impacts have prevented this approach from being adopted as a long-term solution. The fact that these chemicals spread in water makes nontarget impacts particularly likely. The eradication of the Caribbean black-striped mussel (Chapter 9.3) shows that chemicals can be very effective against mollusks, but the degree of nontarget mortality permitted in that campaign would be unacceptable in most freshwater settings, where natural recolonization of the sort that occurred in Darwin Harbor would not be possible.

Chemicals have been used for at least a century to attempt to eradicate small populations of invasive freshwater fishes. Rotenone, for example, was originally a broad-spectrum insecticide but has long been used to kill undesirable fishes in small areas, such as a pond or short reach of a stream. For instance, rotenone was used in New Zealand to eliminate introduced mosquito fish from several small water bodies and has similarly removed the Asian topmouth gudgeon from small lakes in Great Britain. Rotenone is by far the most commonly used piscicide. In addition to rotenone, other piscicides have been widely used for specific species. The sea lamprey invasion into the Great Lakes and other lakes in northeastern North America has long been controlled in a traditional maintenance management scheme with some success through use of the lampricides TFM and niclosamide combined with barriers such as dams; eradication was not a goal. A recent innovation (Chapter 12.4) will probably replace this approach in many areas. From a regional standpoint, the use of piscicides such as rotenone to eradicate local invasive fish populations can be seen as maintenance management, as reintroduction is expected sooner or later so long as the target species remains present regionally. Piscicides are not used in maintenance management the way some insecticides are—by more or less routine, ongoing application over years—because they also kill native fishes and some aquatic invertebrates and because of the expense. Another limitation is that they would be diluted to ineffectiveness in large water bodies.

Many chemicals have been used to kill mammals, both in eradication campaigns and in long-term maintenance management. Although deployed in other ways (for instance, as burrow fumigants against rodents), by far the most common method for using chemicals for mammal management is in toxic baits. These often contain anticoagulants—compounds that inhibit the synthesis of clotting factors. The main technical difficulty with using anticoagulants is that they all have nontarget impacts to a greater or lesser extent, particularly affecting mammals and birds. Warfarin was the first such anticoagulant developed for mammal control, and the evolution of resistance to Warfarin by several targeted rodents led to development of other anticoagulants, such as the widely used brodifacoum and diphacinone. Brodifacoum has been used in many of the successful island rodent eradications described previously, but it is also used for maintenance management of the introduced brushtail possum in New Zealand. Diphacinone is often used in place of brodifacoum where nontarget impacts are a concern. In addition to its use in eradicating rats on some islands, diphacinone is used in maintenance management of invasive rat populations in Hawaii and Puerto Rico.

Other toxicants used in mammal baits are not anticoagulants and work in a variety of ways. Most commonly employed in eradication and maintenance management programs is sodium monofluoroacetate, widely known as 1080. It has been used to manage brushtail possums in New Zealand, feral pigs, red foxes, and cats in Australia, and European rabbits in both countries.

Controversy over the use of both anticoagulants and 1080 (Chapter 11.6) has led in two directions. First is the development of toxicants viewed as more humane and more target-specific, although none of these are yet widely employed. Perhaps most promising is PAPP (para-aminopropiophenone), which affects individuals similarly to carbon monoxide poisoning and has recently been approved for use in managing stoats in New Zealand. The second response to controversy over poisoning mammals is the development of baiting strategies to minimize ingestion of bait by nontarget species. These latter methods can include the physical design of the bait station to make it difficult or impossible for any but the target species to reach it. Baits can be buried so that most nontarget species will not attempt to reach it; this method is used against red foxes in Australia. Baits can also consist of foods that are attractive only to certain species, or they can be combined with various olfactory or visual lures that aim specifically at the target species. However, a certain amount of nontarget mortality is almost always associated with the use of these toxicants when nontarget mammals are present. Also, scavenging of cadavers of target animals killed by toxicants can lead to secondary poisoning of mammals, birds, and even reptiles. Secondary poisoning can be minimized by bait placement and timing, but it can never be totally eliminated if scavengers are present.

Chemical control is plagued by two generic problems, both of which have been mentioned in cited examples. The first is that biological invaders, as do all living organisms, evolve by natural selection to adapt to any impediments to their survival and reproduction, and this includes pesticides and herbicides. Rabbits in Australia have evolved resistance to 1080 (Chapter 5.2), hydrilla in Florida has evolved resistance to fluridone, and the diamondback moth has evolved resistance to several different chemical insecticides as well as to Bt. Resistance can be physiological or behavioral, but it is a constant threat. In response to resistance, managers must either find another method (perhaps another pesticide) or use increasing amounts of whatever chemical they are using to maintain the same degree of mortality of the target species. The latter strategy, in turn, leads to greater expense and often to greater nontarget impacts.

The second generic problem with chemical control is the expense, particularly if the goal is managing large areas for ecological purposes rather than to produce a crop with an evident market value. Herbicides and pesticides are commercial products, invented and manufactured by people aiming to profit from them, above and beyond recovering their development and production costs. Their main interest is thus to sell them at whatever cost the market can bear. Whereas producers of an agricultural crop for which consumer demand is high can often sustain the expense of using large amounts of an expensive chemical (simply by raising prices), government agencies and nongovernmental organizations managing large tracts of land or water bodies for the public good cannot easily follow suit. Public resources for conservation purposes are always limited, and the increased expense of managing an invader through use of a chemical will come at the cost of fewer resources devoted to some other conservation cause.

10.3 What are the advantages and disadvantages of classical biological control?

The expense and nontarget impacts associated with chemical control have led to a search for alternative methods of maintenance management, especially classical biological control—the introduction of natural enemies of the target invader, usually from the invader's own native range. Thus, biological control is akin to fighting fire with fire: introducing a species in the hope of controlling a previously introduced species. The fact that biological control, properly done, does not have the nontarget impacts associated with broad-spectrum pesticides, and especially the fact that it is unlikely to have human health impacts, has led many to see biological control as a green alternative to chemical control. And although the exploration to find potential biological control agents and the testing to determine if they are likely to be effective and not have nontarget impacts can be expensive, once a biological control agent has established a population there should be no further expense. The biological control agent should operate in perpetuity to maintain the population of the target invader at low densities. In theory, there will be a homeostatic relationship, with an increase in the population of the pest automatically engendering an increase in that of the natural enemy, which in turn will cause the pest population to decline, which in turn will cause the population of the natural enemy to decline, and so forth ad infinitum. Biological control is a very popular approach; since 1996, there have been approximately 1,000 deliberate releases of enemies of invasive plants and 5,000 deliberate releases of enemies of invasive arthropods.

The early development of biological control methods primarily served agriculture and rangeland and included some striking successes. During the late 19th century, the vedalia lady beetle from Australia controlled the cottony cushion scale on California citrus, and the cochineal insect from tropical America controlled prickly pear cactus in Sri Lanka. A complex of American insects introduced to Hawaii in the early 20th century succeeded in controlling lantana in drier lowland regions, though it remains a troublesome weed in wetter upland sites. A major success in Australia was the 1925 introduction from Argentina of the cactus moth, which ultimately reduced prickly pear infestations from over 60 million acres to a small fraction of this area today. In the 1940s, klamathweed beetles introduced from Australia to California controlled introduced Klamath weed, which was poisoning livestock.

Such successes are by no means rare, and some of these projects have targeted invaders causing ecological rather than economic problems. Alligatorweed, an aquatic plant from South America that was threatening to choke many of Florida's rivers and lakes, was well controlled in the 1960s by the introduced South American alligatorweed flea beetle (Figure 10.2). More recently, an introduced lady beetle rescued the native gumwood tree on the Atlantic island of St. Helena from attack by the orthezia scale from tropical America (Chapter 3.3).

Despite these triumphs, most biological control projects have not succeeded in controlling the targeted invader. Between 10% and 20% of well-planned projects targeting introduced insects provide some control, and nowadays perhaps 50% of insects introduced to control invasive plants are at least partially successful. It is likely that some of the failures

Figure 10.2 Alligatorweed flea beetle, introduced to control South American alligatorweed in Florida. (Photograph courtesy Ted Center.)

were because the genotypes of the proposed biological control agents that were introduced were not suited to the physical environment of the region where they were introduced. The genetic aspect of the problem is increasingly considered by biological control practitioners, with climate of the source populations matched to that of the location of the target pest with the expectation that the genotypes to be introduced will be suitable for region of introduction.

Some earlier projects have had disastrous conservation consequences by virtue of attacks on nontarget native species. For instance, the small Indian mongoose caused the extinction of native vertebrates after its introduction to several islands for rat control, and the rosy wolf snail eliminated native snails from Pacific islands after its introduction to control the giant African snail (Chapter 3.2). Stoats introduced to New Zealand to control rabbits preyed on native birds (Chapter 3.3). Grass carp introduced to the United States for weed control carried a parasite that ultimately threatened a native fish (Chapter 3.5); in addition, they are not selective feeders and consume native plants as readily as introduced ones. The Old World multi-colored and seven-spot lady beetles, introduced for aphid control, displaced native lady beetles in North America (Chapter 3.3). Both the rosy wolf snail and the mongoose are listed by the International Union for Conservation of Nature (IUCN) as among the world's worst 100 invaders, as are five other species originally introduced for biological control: (1) a flatworm introduced to Pacific islands to control the giant African snail, because it attacks native snails; (2) the cane toad (impacts given in Chapter 3.3); (3) the mosquitofish, which hybridizes with native species (Chapter 3.6) but which also attacks nontarget species as prey; (4) the stoat, just mentioned; and (5) the Indian common myna, which is implicated in the endangerment of several native birds through competition for nest sites, spreads seeds of invasive plants such as lantana, and is a crop pest.

Most of the biological control introductions that have turned out to be conservation disasters happened many years ago, often in the infancy of the technology. Of the 7 biological control species listed among the 100 worst invaders, 4 (the mongoose, cane toad, mosquitofish, and stoat) are generalized predators that attack not only the target species but also many others. For this reason, predatory vertebrates have long been avoided in biological control. The flatworm is also a generalized predator and, though its introduction was more recent (1980), most biological control specialists would strongly advise against introducing such a species today. Oddly, the Food and Agriculture Organization of the United Nations was recommending it as a biological control for the giant African snail as recently as 2002, when its impact was already well-known and deplored. Similarly, today the rosy wolf snail would not be promoted for biological control because of its broad prey spectrum. In general biological control concerns with nontarget impacts are far greater today than they were 50 or more years ago, particularly in projects to control introduced plants.

Nevertheless, two more recent biological control introductions are raising alarms among conservationists. A flowerhead weevil was introduced from France by Agriculture Canada in 1968 to control musk thistle and then was widely distributed by federal and state agencies in the United States. This weevil eats seeds of at least 22 native thistle species, including several of conservation concern, and is the main threat to the Suisun thistle, listed under the federal Endangered Species Act. In this instance, the Canadian project leaders recognized that nontarget impacts might occur, but they tested only 1 native American thistle species (of 88 native species in the same genus as musk thistle) plus 5 European thistles, far too few to assess possible nontarget impacts. In addition, they incorrectly concluded that the fact that the flowerhead weevil selected musk thistle over others for feeding and laying its eggs and grew better on musk thistle ensured that its impact on native thistles would be minor at most. In fact, the presence of large populations of musk thistle could lead to large populations of weevils, which could then devastate a smaller population of native thistles—a classic indirect effect (Chapter 4.1). In addition, if the preferred host thistle is not present, the weevil is likely to attack other thistles even though musk thistle would have been preferred.

A current problem is the spread of the cactus moth in the southern United States. In Australia, where it was so successful at controlling prickly pear, no native host plants are present. The United States and Mexico, however, do contain such host plants. The moth was introduced in 1957 to Nevis in the Lesser Antilles by the Commonwealth Institute of Biological Control to control pest native prickly pear. The moth flew or was blown northward to the Greater Antilles and then reached Florida of its own accord from Cuba or in cut flowers shipped from the Dominican Republic. It has so far spread west to Louisiana and north to South Carolina. Approximately 79 native prickly pear species are at risk in the United States and Mexico as suitable hosts for the cactus moth, and some are important food or ornamental plants.

The experience of a century of biological control introductions and careful consideration of cases like those of the cactus moth and the flowerhead weevil have led to protocols that lower the probability of harmful nontarget effects. They are better developed for insects and pathogens considered for biological control of introduced plants than for enemies of introduced insects. The main approach in testing a potential plant-eating insect is called the centrifugal phylogeny method, formalized in the 1970s, and consists first of testing whether the insect will attack, survive, and reproduce on other varieties of the target plant, then other species of the same genus, then other species of the same subfamily, then other species of the same family. If a sufficient number of host species are tested, it should become clear to what degree the potential biological control agent will attack nontarget species and perhaps affect their populations. Other plant species of particular conservation or economic significance should be tested in addition. For instance, in tests for biological control agents for introduced pest thistles, tests on artichokes would be expected, no matter how distantly related to the target pest they are, because artichokes are a thistle of great economic value. Nowadays other plants are sometimes added to the test list if they are known to be morphologically or chemically similar to the target pest plant, even if they are distantly related.

An important point is to use no-choice tests—that is, to present each test species to the proposed biological control insect without giving it a choice of other plants to eat or to lay eggs on. This is because, as noted for the flowerhead weevil, an introduced insect can devastate a plant of conservation concern even if the latter is not a preferred host. The centrifugal phylogeny method, if rigorously applied, should minimize unexpected nontarget impacts, but it will not completely eliminate them. There are simply insufficient resources to test all possible nontarget species (for instance, all 88 North American thistle species in the same genus as musk thistle); some potential host species could probably not even be collected or reared in a greenhouse.

For biological control of insects, no rigorous analog of the centrifugal phylogeny method is in use. Partly this is because regulations of insect introductions for insect control are far less stringent than those governing introductions for plant control. It is also partly because of technical difficulties. The evolutionary relationships of many insect groups are not well-known, so it would be difficult to arrange them in centrifugal array of species increasingly distantly related to the target species. Also, whereas plant-feeding insects have been found to be more likely to feed on near relatives of a known host plant than on distantly related species, some insects parasitic on other insects attack distantly related hosts so long as they occupy the same habitats. And a more general problem with insects is that there are so many of them, and so many are poorly known. About 265,000 plant species are known, and it is estimated that perhaps 320,000 exist total. For insects, about 760,000 species are known, and this is a much smaller fraction of the total believed to exist—perhaps 8 million species. So one might well not be able to find most of the species closely related to the target host that ought to be tested. One might not even know if they exist. Another problem is that rearing plants in a greenhouse for test purposes is difficult enough. To rear all the host plants that would be needed to rear all the insects that would have to be tested if the biological control project targeted an insect would be a Herculean task.

This is not to say that insect biological control projects do not test for possible nontarget impacts. They do, but the testing is not very systematic and is often rather haphazard. Consider biological control projects aimed at managing two introduced insects currently wreaking havoc in North American forests: the hemlock woolly adelgid (Chapter 3.1) and the emerald ash borer (Chapter 6.6). In both instances, forests are being destroyed rapidly and chemical control is feasible only very locally, so it is unsurprising that biological control should be attempted. For the hemlock woolly adelgid, two Asian beetles have been released: the hemlock woolly adelgid lady beetle and a tiny black beetle known, appropriately, as the tiny black beetle. For the lady beetle, fewer than six possible other host species were tested (of which the beetle also fed and completed development on three). There are surely at least 100 possible native aphid, adelgid, and similar native hosts. The crisis was deemed so urgent that 100,000 of these beetles were released in Connecticut while they were still undergoing evaluation for nontarget impacts. The tiny black beetle was tested on three adelgid species in addition to the hemlock woolly adelgid, two aphid species, and a scale insect. It ate the eggs of all the other species and laid its eggs on the other adelgid species. There is no evidence yet that either of these beetles has impacted the spread of the hemlock woolly adelgid or decreased its impact on hemlocks (or that either has affected nontarget insects), but the project continues, in the hope that future releases of these two beetles, or others yet to be found, will slow down this scourge.

The emerald ash borer presents a greater challenge in terms of nontarget impacts, as it is a jewel beetle. These are prized by collectors, and at least 175 species are in the same genus as the emerald ash borer in the United States and hundreds more in Mexico. Many of these are rare and very poorly known, so testing a biological control insect on them in the laboratory would be impossible even if massive resources were available; in fact, only nine species were tested as possible hosts. Three parasitic wasps have been released in the United States to combat the emerald ash borer. Only one of these rejected the eight other beetle species and laid its eggs only on the emerald ash borer. At least one of the other two attacked nontarget beetle species in significant numbers. Two of these wasps have established populations and are spreading. As with the beetle releases for the hemlock woolly adelgid, there is no evidence that these wasps have mitigated the emerald ash borer invasion, but hope is high that at least some trees will survive and that current seedlings will replace some of the lost ash forests. There is certainly no evidence that these wasps have affected populations of any nontarget beetle. However, even if they did, it would be unlikely we would know this for decades, if ever. There is little if any monitoring of jewel beetle populations.

The spread of the cactus moth from the West Indies to Florida and beyond is one of several cases that have led many specialists to ponder possible movement to unintended areas when considering potential biological control introductions. Such movement can occur either autonomously, as all species have adaptations for some sort of dispersal, or with human assistance, either deliberate or inadvertent. For instance, the introduction to Tahiti of a wasp that parasitizes eggs of an insect plant pest, the glassy-winged sharpshooter, led to the accidental introduction of the wasp to many other islands in French Polynesia, some as far as 850 miles from Tahiti, as a contaminant of sharpshooter-infested plant material. Enough cases are known of deliberate movement of species not intended to be introduced beyond a certain point that one can question the wisdom of certain introductions. A recent prominent case involved the release of the northern tamarisk beetle from Asia to control salt cedar in Utah. However, tamarisk (salt cedar) has so dominated parts of the Southwest that it has become a prominent nesting habitat for the southwestern willow flycatcher, a bird on the Federal Endangered Species list. Concerned that devastation of salt cedar by the beetle would lower the bird population, the USDA restricted beetle releases to sites more than 200 miles from known flycatcher nests in salt cedar. Nevertheless, in the wake of much publicity about the possibility that the beetle could reduce salt cedar infestations and campaigns by Utah and local governments to spread the beetle more widely, individuals simply carried the beetle much farther south into the range of the flycatcher. The USDA then canceled the release program, although the beetle is already established. The possibility of such rogue introductions of biological control agents by persons eager to attack some invader will always be present, and sometimes people will go to great lengths to do this. Rabbit calicivirus being tested on Wardang Island off the coast of South Australia arrived by unknown means not only on nearby coastal mainland but also hundreds of miles inland, and in 1997 it was illegally introduced to New Zealand and dispersed by farmers.

In sum, because biological control agents are living organisms, their introduction will always entail an element of risk. All living organisms have some sort of autonomous dispersal mechanism, and we cannot tell exactly where they will go or be carried. All living organisms also reproduce, so they will not simply go away eventually if undesired side effects arise, as a pesticide or herbicide might dissipate. Often it takes very few individuals to start a new population. All living organisms evolve, and we cannot predict exactly how they will evolve or how quickly. A biological control agent that originally attacked only the target pest could evolve a taste for a nontarget host. Many plant-eating insects have evolved tastes for new host plants; examples in Chapter 5.2 depicted how several, such as the Colorado potato beetle, became worse pests. Similarly, insect parasites of other insects evolve to use new hosts. So far no striking example is known for biological control insects in which evolution of new host or prey use has caused an important nontarget impact (although, for insects attacking other insects, such phenomena might occur and our general lack of knowledge would prevent our knowing about it). Is this just luck? It is impossible to tell today.

The full nontarget impact of some biological control introductions against insect invaders might well be enormous but will probably never be completely understood. The tachinid fly introduced in an attempt to control the gypsy moth fortuitously ended up controlling the browntail moth (Chapter 4.4). However, the story is far more complicated. This fly was introduced several times between 1906 and 1986 against 13 different pest species. For some it became a substantial factor; for others it did not. For some pestiferous nontarget invaders, such as the browntail moth and the imported cabbage worm, it provided unexpected control, and it also affected some native agricultural pests, such as the cabbage looper. However, this tachinid fly is known to affect giant silk moths, including some of conservation concern. In studies in New England, it accounted for 81% of the deaths of the cecropia moth, 68% of the deaths of the promethea moth, and 36% of the deaths of the buck moth. There is every reason to think its impact on some less obvious insects is just as great. A parasitic wasp that specializes on aphids has a similarly wide but poorly studied impact in Europe. Its biology and impact are so poorly known that it does not even have a common name. It was introduced from Cuba to Czechoslovakia in 1973 for study as a potential control of two nonnative aphid pests of citrus in Europe. Liberated in southeastern France and the island of Corsica in 1973, by 1986 it had spread throughout Mediterranean Europe and was the main parasitic wasp attacking both native and introduced aphids, at least 26 species, both pests and innocuous ones, in many habitats, from agricultural fields to forests. The multicolored lady beetle and the seven-spot lady beetle are by far the most common lady beetles in some parts of the United States now (Chapter 3), displacing native lady beetles and eating many aphids and other small, soft-bodied insects, both introduced and native, both pestiferous and harmless species. Their full ecosystem impact can only be guessed.

10.4 What is augmentative biological control?

In classical biological control, the introduced biological control agent and the target pest continue to act in perpetuity in a homeostatic fashion, without further human intervention. When the pest population grows, the control agent population grows and brings about a reduction in the pest population. However, for certain introduced pest species, even if the introduced biological control agent does attack the target pest, it does not consistently keep the pest population at a sufficiently low level. In such circumstances, augmentative biological control has occasionally proven effective. In this method, populations of an introduced (or sometimes native) enemy of the target pest are produced, often in large mass-rearing facilities, and are periodically distributed into the field. Thus, the reproduction and spread of the natural enemy do not have to "keep up" on their own with the reproduction and spread of the pest. Instead, humans help them.

An excellent example of augmentative biological control is the management of the sirex woodwasp in Australia and Brazil. The sirex woodwasp is native to parts of Eurasia and north Africa and reached Australia in the 1950s and South America in the 1980s. It attacks many species of pine trees. Pine trees are native in the northern hemisphere, but they are introduced in Australia and South America, where they are the basis of a massive timber industry. This industry was threatened by the sirex woodwasp, which can kill a tree in just a year. The threat led the Australians to a research effort, led by zoologist Robin Bedding, that culminated in a remarkably effective means of managing the wasp by augmentative biological control. The woodwasp carries around a symbiotic fungus that females inject into trees when they lay their eggs; it is the rapid growth of this fungus that causes the lethal wilt of the attacked trees. Bedding discovered a roundworm, native to Australia, whose biology is remarkably suited to that of the woodwasp. This roundworm has two forms, one of which feeds (and rapidly reproduces) on the fungus. If, during their feeding, these fungus-eating roundworms detect woodwasp larvae, they develop into a parasitic form that enters the woodwasp larvae and produces offspring, which migrate into the reproductive organs of the developing woodwasp and eventually into the eggs. When the parasitized female woodwasp then lays her eggs in a pine tree, she injects some of the fungus, and her eggs hatch into nematodes instead of woodwasps. The nematodes start eating the fungus and the entire cycle begins again.

However, over time the woodwasp population eventually escapes adequate control by the nematode if the nematode is left simply to reproduce and disperse on its own in nature. In Australia and Brazil, the fungus is cultured in large rearing facilities, and millions of nematodes are cultured on the fungus, collected, and then distributed in the field. This method is so effective and valuable that rearing nematodes is a commercial enterprise in Australia. A Brazilian government agency has introduced the nematode from Australia, built massive culturing facilities to produce the fungus and the nematode, and forestry companies employ large teams of workers to distribute the nematode over pine plantations in continuous fashion.

10.5 What is mating disruption?

Several introduced insect populations have been managed at least partly by a technique known as *mating disruption*, in which the chemical signals that males and females use to find one another to mate are turned against them. The method generally entails first identifying the chemical, or sex *pheromone*, that females emit to attract males, then synthesizing a version of it and releasing it in huge amounts. Males are sensitive to the female sex pheromone in infinitesimal amounts and often during only a short period during the day. Continuous exposure to massive amounts of the pheromone may desensitize the males to its odor. Also, when the pheromone is present throughout an area in great amounts, the males may have a difficult time locating a female emitting normal, minute amounts. In fact, males often fly to the apparatus dispensing the synthetic pheromone.

This method has been used against many moth species, often quite successfully, and it has been attempted against a few other introduced insects, such as mealybugs and scale insects. For example, mating disruption is used widely in North America against the codling moth, a European species that is a major pest of apples, pears, and walnuts. Many tons of the active ingredient of the pheromone are synthesized and dispersed over more than 200,000 acres in the United States. If the population density of overwintering codling moth larvae in orchards is very great (usually over 400 per acre), mating disruption is supplemented with an insecticide.

An ingenious possible extension of mating disruption, called *mobile mating disruption*, was recently proposed by entomologist Max Suckling and his colleagues, and a pilot field test produced promising results. Noting that, in California, occasional outbreaks of the medfly (Chapter 9.4) are eradicated by massive release of sterile males (Chapter 9.3), they considered the possibility of using these sterile male medflies to deal with a different pest. The light brown apple moth (LBAM; Chapter 9.4), a recently invading agricultural pest in California, was being subjected to airplane sprays of its female sex pheromone, but public opposition to the widespread spraying forced a halt to this program. Suckling and his coworkers instead dosed the sterile medflies with a solution of the LBAM pheromone, determining that this procedure did not harm the flies. The flies, carrying the pheromone, were then released in great numbers in a small area and buzzed around the region confounding the LBAM males about the location of LBAM females. The method must now be tested on a wide scale. Because the California public has not protested the frequent release of sterile male medflies, using them to disrupt the mating of the light brown apple moth would not engender the opposition aroused by aerial sprays.

10.6 What is integrated pest management?

The many nontarget impacts of chemicals as well as problems with evolution of resistance and expense have led to the development of a method known as integrated pest management

(IPM) for control of introduced and native pests of agriculture. The period of the late 1940s through the mid-1960s is sometimes called the dark ages of pest control because of the increasing reliance on insecticides. The impetus for IPM was the observation toward the end of this period that routine use of insecticides sometimes worsened a pest problem, because the insecticide killed natural enemies of the target pest more effectively than it killed the pest itself. Evolution of resistance by the pest sometimes exacerbated the situation, as did population surges of new pests that had previously been suppressed by natural enemies. The human health and ecological impacts of chemicals used as insecticides, publicized by Rachel Carson, also spurred a search for alternative approaches. Among precipitating events was the spread of the Asian pink bollworm, a cotton pest, into Arizona and California, causing chemical sprays, especially DDT, to be used up to 25 times annually. Residues of DDT began to appear in milk at concentrations surpassing federal tolerances. Also, outbreaks of the gypsy moth could not be controlled in the absence of DDT, banned since 1957 for use in forests.

In response to this crisis, several entomologists in the mid-1960s developed the foundations for what they first termed integrated control, later called integrated pest management. Unfortunately, this term has by now acquired so many definitions (at least 70 by recent count) that it has become a panchreston—a term that means so many different things that it means almost nothing. An early project sponsored by U.S. federal agencies aimed to reduce insecticide use by 40–50% in five years on six crops by careful study of pest life cycles. The key innovations were to time chemical sprays so that they would be used only at the times of greatest insect vulnerability and to restrict them only to parts of the landscape where the pest concentrations were high—thus avoiding adjacent habitats where natural enemies might reside and incidentally slowing the rate at which resistance evolves. Other projects subsequently incorporated other features, such as the use of sterile males and females to hinder mating by the codling moth, an introduced pest of apples.

Some versions of IPM also incorporate the use of biological control, through the careful use of insecticides to avoid devastating the control agents, especially by limited timing and location of spraying. Management of the habitat, through such practices as crop rotation and thoughtful adjustment of harvest and irrigation schedules, can also sometimes be used to aid beneficial insects and disfavor pests. However, IPM as implemented often consists of little more than using insecticides, but judiciously. One common formulation calls for the use of economic thresholds—that is, the recognition of a density of some target pest such that, below that level, the economic cost of the damage to the crop can be tolerated. The insecticide is then not used unless pest density exceeds the economic threshold. Although in principle such a scheme could lessen insecticide use, two aspects are troublesome. First, the threshold can be arbitrarily set according to economic concerns; it can vary, and it could, for example, be set so low that spraying would be the usual management tool. Second, it is reactive rather than proactive. By the time a designated threshold density is reached, the insect population has already increased and a certain amount of damage is insured. A more ecological approach would require a method to keep a pest population from ever reaching the threshold density.

10.7 What is ecosystem management?

Mechanical, chemical, and biological control as well as IPM all target a single invasive species at a time. A method may do double duty—take, for example, the tachinid fly introduced to control gypsy moth that incidentally controlled the browntail moth. Or a chemical that kills one invasive pest may also kill a second. However, the conception of the method is generally to attack a particular species. But introduced invaders often occur together. For instance, a database of over 1,000 conservation projects undertaken by the Nature Conservancy and its partner organizations showed that invasions were the single most common problem and that 69% of all projects with invasive plant problems reported two or more problematic invasive species in a single habitat. Some reported as many as 12. This sort of observation, and the attraction of possible economies of scale, led to great interest in the concept of ecosystem management as a tool to combat invaders. Ecosystem management is the management of entire ecosystems so as to favor native species in general and to inhibit introduced species. Ecosystem management gained rapid adherents in many branches of resource management in the 1990s, as single-species management for various purposes seemed at best inefficient and at worst a failure. Many management issues in addition to invasions spurred this enthusiasm. For instance, the use of a single species by the USDA Forest Service as an indicator species of the health of entire forest communities (both plants and animals) led to grave concerns that the status of no single species could say enough about the status of the entire ecosystem and its many species. In addition, managing an indicator species to improve its status would not automatically improve the condition of the forest community whose status it was supposed to indicate. For example, many tactics employed to increase populations of the red-cockaded woodpecker, a specialist and designated indicator of longleaf pine forests in the southeastern United States, were so highly specific that it is hard to imagine they would aid other species—moving individual woodpeckers to new areas, constructing nest cavities for them, and installing devices in nest entrances to exclude snakes.

By the late 1990s, all U.S. federal agencies had adopted ecosystem management as the official guiding principle for managing natural resources, though there was not a single accepted definition. The unifying feature of the various definitions was a focus on managing ecosystem processes, fire and hydrological regimes in particular, rather than on individual species. Integrated pest management had earlier sometimes incorporated managing entire agricultural landscapes to lessen invader impacts, but the key feature continued to be use of chemicals. The emphasis of the advocates of ecosystem management was on processes, with chemicals playing little or no role, and on nonagricultural landscapes such as forests or prairies.

Habitats disturbed in various ways are often more readily invaded than intact habitats (Chapter 2.6), and sometimes a particular invader can modify an ecosystem process, like a fire regime or a nutrient cycle, in such a way as to facilitate invasions by other species. These observations suggest that intelligent management of the same ecosystem processes could inhibit invasions. Furthermore, in some instances, it has long been known that interfering with natural processes fosters certain invasions. For instance, overgrazed pasture is very prone to invasion by musk thistle and other pasture weeds, and prevention of overgrazing greatly reduces this problem. Similarly, fire exclusion often leads to the invasion of prairie by woody species. Changed hydrology is the main driving force for the invasion of south Florida by paperbark and Brazilian pepper. As yet another example, the longleaf pine and wiregrass community (Chapter 2.6) has remarkably few invaders, even though in many areas it is close to large stands of introduced ornamental plants that would seem ideally suited to invade the ground cover. Even the red imported fire ant, so prominent in disturbed habitats throughout the range of longleaf pine, is not found in intact forests except along roads. As this ecosystem is maintained by frequent, low-intensity, naturally occurring fires, a hypothesis for the lack of invasion is that such a fire regime is simply not compatible with the existence of many introduced species.

There seems little doubt that careful management of entire ecosystems to retain important features of naturally occurring ecosystems would lessen the magnitude of invasion, though three factors suggest this is not likely to be a magic-bullet solution that will obviate the need for good single-species management. First, resources are increasingly extracted from all manner of ecosystems, and they are modified in many other ways as well—for instance, by housing and road construction. Short of ceasing such activity, there is a limit to how much we can make managed ecosystems resemble pristine ones. Second, it takes an enormous amount of research to understand exactly which features of intact ecosystems favor native species and disfavor invasive ones. For most ecosystems, this research has not been conducted. Finally, even intact ecosystems are occasionally invaded, just less frequently than disturbed ones. No matter what the set of conditions maintaining an intact system, some nonnative species will be able to thrive in it. For fire-maintained longleaf pine forests, Asian cogongrass is just such a threat (Chapter 2.6) and in fact has begun to invade in some sites.

In short, there is no foolproof way to stop invaders from disrupting an ecosystem. Ecosystems comprise many different interdependent organisms and processes, so an invader can disrupt them in myriad ways. It is therefore important that management projects use a number of strategies, both general and specific, to mitigate invasion impacts.

11

CONTROVERSIES
SURROUNDING BIOLOGICAL
INVASIONS

11.1 Which introduced species are harmful and which are useful?

Several critics suggest that rather few introduced species are harmful and some are even useful, so the angst experienced over invasions and the publicity generated about them are overblown. It is true that most introduced species are not known to be ecologically or economically harmful, and it is also true that some are useful; nevertheless, it is illogical to argue that we should not worry so much about them. After all, most bacteria and viruses are not known to harm human health or other interests, and some are even useful (for instance, bacteria that help plants fix nitrogen or farmers make cheese, viruses that kill insect pests). Yet we worry about microbes in general and publicize human diseases caused by pathogens.

The problems with the general argument that invasion impacts are overstated begin with the assertion that few introduced species are harmful. We do not know of harmful impacts of most established introduced species, but we know that many do have harmful impacts. Chapters 3–5 outline the variety of impacts, but new ones are detected every year. We also know that some of these impacts did not happen for a long time and that others were sufficiently subtle that, even though they were occurring, they were not recognized for a long time. In fact, both policy and management efforts are focused on impacts, not simply introductions. Article 8h of the Rio Convention on Biological Diversity (Chapter 8.1) calls for interdiction, eradication, and management not of all species but of "those alien species which threaten ecosystems, habitats or species." We have learned in the 20 years since Rio that far more invasions may pose threats than was realized in 1993, but the focus remains on potential and realized impacts, not on the foreign origin of a population per se.

Though native species sometimes inflict the same sorts of damage that introduced ones do, they are much less likely to do so. A recent study shows that, in the United States, introduced plant species are 40 times more likely than native ones to cause ecological harm. When native species do become damaging, this is almost always precipitated by human activity. For instance, in North America several native juniper species have encroached on prairie or grassland in the wake of intensive grazing by livestock and suppression of naturally occurring fires. When native species cause damage, we try to curtail them just as with introduced species. Native deer, for instance, occasionally become so abundant in the absence of predators that are long gone or now very rare, such as wolves and mountain lions in the United States, that they devastate vegetation. In such instances, hunting is encouraged, and sometimes commercial hunters are contracted to reduce herds.

It is also true that some introduced species are useful from certain perspectives. Most leading food crops in the United States are introduced (Chapter 1.4). Some introduced game fish, birds, and mammals are prized by hunters and fishermen, who would surely consider them useful, although some of these species have impacts on native species that would be deplored by other stakeholders. For instance, introduced

rainbow trout from North America and brown trout from
Europe thrive in New Zealand waters and attract fishermen
from all over the world. However, these same trout have
greatly reduced populations of native fish by both predation
and competition. In some streams, these trout consume the
entire annual production of insects, initiating a trophic cascade (Chapter 4.1) that includes great growth of aquatic plants
in the absence of insect herbivores—in short, a transformation
of the entire ecosystem.

Introduced species can occasionally even aid conservation of a threatened species or dwindling natural habitat.
Restoration ecologists have long recognized that some plant
species, even plants that would be damaging invaders in other
circumstances, can act temporarily as nurse plants in restoring
various plant communities, especially forests. For instance,
introduced Caribbean pine, which is a damaging invader in
New Caledonia, facilitated the reestablishment in Sri Lanka of
several native tropical tree species that would have colonized
a degraded site much later, if at all, without pine. In other situations, an introduced near relative may functionally replace
extinct species. For instance, preliminary studies on two small
islands off the coast of Mauritius where huge native tortoises
are long extinct suggest the Aldabra giant tortoise will serve
this purpose. Even as generally destructive an invader as the
ship rat pollinates some native plant species in New Zealand,
replacing locally extirpated birds. The fact that the ship rat
was instrumental in eliminating the birds in the first place
shows that, even though the ship rat has a beneficial use now
(along with a plethora of harmful effects), it would have been
far better had the invasion never occurred.

11.2 How do introduced species affect biodiversity?

Ecologist Dov Sax and colleagues point out that, in several
locations, establishment of introduced species has equaled
or even surpassed extinction of native species, so that

biodiversity is unchanged or even increased. For instance, for bird species on oceanic islands worldwide, the number of species has tended to remain approximately the same, even as massive numbers of native species have been extinguished by human impacts (hunting and habitat change) and introduced predators. On many islands, there have been so many introductions of bird species from elsewhere that they have counterbalanced, in some sense, the extinctions. For plant species on oceanic islands, the data are not as complete or reliable, but the pattern seems even to be exaggerated—there have on average been more invasions than extinctions. At some continental sites, a similar pattern holds. For instance, the number of plant species in individual states of the United States has increased by an average of about 20% in the last two centuries, with massive introductions and rather few extinctions. Similarly, freshwater fish diversity has increased in many river drainages of the United States. As Sax and his coauthors acknowledge, at the same time as these local increases in numbers of species occur, there are global decreases in all these species groups.

It is not so clear that local increases in species diversity should be viewed as a benefit. Nature writer David Quammen evokes a future in which the earth is a "planet of weeds." He predicts that, because of introductions and extinctions, in the future most places will still have many species, but they will tend to be the same species everywhere: the global weeds, both plants and animals. The earth as a whole will have fewer species. In his view, humankind will be impoverished by this change.

We should consider the birds of the Hawaiian Islands in this context. Hawaiian native birds evolved after about 20 separate arrivals of birds in the prehistoric past from continents or other islands, mostly species associated with water. Because of the tremendous distances, these natural immigrations occurred only at intervals of hundreds of thousands of years. The isolation of the archipelago that so restricted immigration also ensured that the few survivors would evolve to differ from their forbears; over 80% of native Hawaiian birds are found only in Hawaii. The best-known group is the honeycreepers, descended from one South American seed-eating finch species that managed to reach Kauai (which was then the youngest and largest of the islands) between 3 and 4 million years ago in numbers great enough to establish a population. Descendants of this population dispersed to other islands and spread farther back and forth among the various islands, leading to the evolution of at least 56 species (of which only 23 still exist) that occupy ecological niches that insect-eaters, nectar-feeders, and other seed-eating birds would fill on continents. Altogether, at least 114 native bird species are known to have lived in the Hawaiian archipelago before humans arrived.

Most people would consider it a misfortune that the majority (at least 66) of these bird species disappeared in a hecatomb beginning about a millennium ago with the arrival of Polynesians, the first humans to reach Hawaii. Introduced species played a leading role in the demise of the many of the birds. Pacific rats hitchhiked with the Polynesians and quickly destroyed lowland palm forests by gnawing on palm seeds. Several native bird species adapted to forest habitat disappeared as forest was replaced by grasses and shrubs, in some cases long before Polynesians established settlements in the region. Polynesians brought pigs, and rooting by pigs also changed plant communities. And of course, the Polynesians hunted birds. Among extinct species known from fossils from this period are several flightless birds, including geese, rails, and ibises. Three owl species, two crows, and hawk were also lost. Most of the 33 extinct honeycreepers disappeared between the arrival of Polynesians and colonization by Europeans and North Americans in the 19th century. This later colonization only exacerbated the loss of native birds. The southern house mosquito stowed away in water casks on a whaling ship from Mexico in 1826. Then in the early 20th century, when acclimatization societies introduced birds from all over the world (Chapter 6.3), several of these from Asia turned out to be excellent reservoirs for the plasmodium that causes avian malaria as well as the virus that causes avian pox—these diseases were vectored by the mosquito. The introduced birds were immune, having coevolved with these pathogens for millions of years. The native birds, however, were devastated (Chapter 3.5). The North American mallard, introduced for hunting, poses a different threat. It hybridizes with the native duck, the koloa, to the extent that on Oahu and Maui no purebred koloas remain.

Ironically, Hawaii also touts itself as a bird-watcher's paradise, and in one sense this is true. The remaining native birds are largely restricted to relatively inaccessible mountain habitats, but 53 introduced bird species from all over the world have established populations. On the campus of the University of Hawaii in a Honolulu suburb, a bird-watcher can see species from Asia, Europe, Africa, and North and South America—northern cardinals, red-crested cardinals, Indian white-rumped shamas, Japanese white-eyes, red-vented bulbuls, house sparrows, house finches, common waxbills, zebra doves. It is exciting to watch them and pleasant to hear their songs in this ornithological United Nations. However, aside from vectoring avian malaria and avian pox and hybridizing with the native duck, the introduced birds of Hawaii are key players in several invasional meltdowns (Chapter 4.2). Common mynas disperse seeds of lantana, which now dominates many upland pastures. Japanese white-eyes disperse seeds of firebush, the nitrogen fixer that is transforming forests (Chapter 3.1).

Should we rejoice that the 66 extinctions are almost balanced by the 53 introductions? Do the latter compensate in some sense for the former? For birds, at least, Hawaii seems to be a forerunner of the planet of weeds envisioned by Quammen. The introduced birds of Hawaii are almost all common in their native ranges, and many are common invaders elsewhere. The birds that have been lost were almost all restricted to Hawaii and will never be seen again. People can still watch birds in Hawaii—lots of them. The loss of more than half of the native bird species does not seem by itself to have caused natural or human-constructed ecosystems to malfunction or to have generated other problems that humans worry about. Perhaps many people would not view this exchange as a tragedy. Others would see it as a terrible loss. Philosophers such as Holmes Rolston III and Lawrence E. Johnson believe we have an ethical duty to save species above and beyond any considerations owed to individuals of those species. Other philosophers, such as Gary E. Varner, believe that only individuals, not collective entities such as species, have rights. Certainly those who see species as what philosophers term *morally considerable* would deplore the extinction of native species even if just as many others or even more took their place. The matter of the rights of species will be discussed later in this chapter with respect to controversies over management methods.

If we do worry about extinction, whether or not numbers are locally redressed by invaders, there is little doubt that biological invasions are a leading cause. Many examples of invasion-driven extinctions are provided in Chapters 1–4, and these are but a small sample of those that are known. Many groups of species (for example, insects) are so poorly known that, if a species disappeared, we would be unlikely even to know about it with certainty, much less to know why it happened. For groups that are better known, information on the role of invasions in extinction is telling. Birds, for example, are far better known than other large groups, partly because they are often so highly visible and partly because they have engaged a large public constituency—bird-watchers. Of 1,186 bird species (12% of the total global number) currently threatened with extinction, almost half (510) are threatened wholly or in part by introduced species—introduced predators, some described in detail in earlier chapters, threaten 298 of them, but 72 are threatened by introduced competitors (including other birds) and 140 by habitat change caused by introduced plants and herbivores.

It could also be argued that extinction is not the only or even the best currency with which to evaluate the impact of invasions. Extinctions are straightforward to tabulate where the data exist, but invasions have many other enormous impacts (Chapters 3 and 4), some as irreversible in practical terms as extinction is. Vast swaths of the earth are now utterly dominated by introduced species, to the near exclusion of native species and the ecosystems that originally existed. In south Florida, for instance, about 700,000 acres that had been the famous river of grass, dominated by native sawgrass and muhly grass, are now forests of Brazilian pepper and Australian paperbark (Chapter 3.1). To my knowledge, no species are yet extinct because of this change (though some are threatened). Does that mean no harm was done? Or should we lament the loss of this vast extent of a native ecosystem? Certainly society as a whole seems to view it as a tragedy, to be stemmed and if possible partly to be reversed. This is reflected by frequent promises by agencies of the U.S. federal government and that of Florida to "save the Everglades" and possibly to restore parts of it, at great expense. But are there good reasons, other than widespread nostalgia, to worry about this change so long as no species disappear completely? In south Florida, the destruction of the grass-dominated prairies also contributes to practical problems for humans, particularly because of changes to the hydrological regime that affect water quality and commercial fisheries. However, suppose there were no such ecosystem services to worry about? Should we mourn the loss of the river of grass? Should we try to rescue it?

It is also likely that there is an extinction debt—that is, a roster of species doomed to extinction by combined forces, including invasive species, that is gradually reducing their population sizes and areas of occupancy. An introduced predator such as a mongoose or rat on an island can drive a native bird to extinction in a remarkably short time—decades. But other forces take much longer to achieve this end. We know from paleontological data that many species that were once very common and widely distributed are now extinct. Some of these losses we attribute to changes in the physical environment and some to sudden events like meteorites. We attribute others to competition, and these took much longer. For instance, the establishment of the Panamanian land bridge 3 million years ago led to the Great American Interchange, with species of North America moving into South America and vice versa. The most striking result was the extinction of many South American marsupial mammals, believed for the most part to have been largely caused by their being outcompeted by North American placental mammals. However, these extinctions took a long time—millennia, often many millennia. Many species today, on continents as well as islands, are dwindling toward extinction, even if the final death throes will be centuries from now. The decline of a substantial fraction of those is due wholly or partly to introduced species.

11.3 How do we know a species is introduced and not native?

As was pointed out in Chapter 1, invasion biologists focus primarily on invasions that have occurred in the last 500 years, even though some invasions, such as that of the Pacific rat to many Pacific islands, occurred well before then. What evidence is brought to bear on the question of whether a species is native to a region or not? The European explorers who traveled to the New World beginning 500 years ago often brought back specimens of plants and animals that they found there; indeed, beginning with Captain Cook's voyages in the late 18th century, expeditions frequently had naturalists onboard specifically to determine which species were present in newly explored sites. Museum collections have countless thousands of specimens with associated data on site and collection date, and these records can also help show whether a species is native.

Often, however, museum specimens and written reports cannot suffice to determine the status of a species. In such cases, indirect, circumstantial evidence can come into play— careful consideration of the ecology of a species, interactions with coexisting species, and, increasingly, molecular genetic information. The common periwinkle, one of the most prominent snails on rocky shores of the European and North American Atlantic coast, exemplifies the process (Figure 11.1). The first North American historical record was in 1840 from a site in Nova Scotia on the Gulf of Saint Lawrence, and the second report from that area was from 1855; this was long after all the other large mollusks of the region had been repeatedly recorded. In 1863, one author considered this snail native based on testimony of elderly Nova Scotia residents that they had often picked up shells like the periwinkle around Halifax as schoolboys. Others questioned whether such ancient memories of nonexperts could be trusted, pointing out that the native flat periwinkle and rough periwinkle can be mistaken

Figure 11.1 European common periwinkle in New England. (Photograph courtesy James Carlton.)

for the common periwinkle. The fact that no living common periwinkles are found in Iceland or Greenland also seemed to point to an introduced status in Nova Scotia. That no shells of the species had been found in middens of Native Americans from Maine and New Brunswick, even though shells of many other species, including the flat periwinkle, were abundant also suggested an introduction had occurred. Given prevailing currents in the opposite direction, it also seemed unlikely that the common periwinkle could have crossed the Atlantic from Europe on its own. The fact that the common periwinkle has long been a popular food in Europe and that European colonists introduced many edible species to North America suggested a ready explanation for its introduction in the early 19th century. It could also have arrived in ballast stones and gravel, as have other European species.

The common periwinkle rapidly spread south after its 19th-century discovery and after about 30 years was the most common snail on rocky shores between Maine and New Jersey (Chapter 3.1). This rapid spread has all the hallmarks of an invasion, as no obvious physical environmental change was recognized that might have caused a native species suddenly to expand its range so dramatically, and no other species spread at that time. However, its introduced status was thrown into doubt when radiocarbon dates from common periwinkle shells found in two Micmac Native American camps in Nova Scotia were reported as 700 ± 225 years, well before European settlement. Based on these dates, one author hypothesized that the common periwinkle arrived on its own about 1,000 years ago on driftwood that traveled from Europe to North Africa, then across the Atlantic near the Equator, and then north to Nova Scotia or New Brunswick in the Gulf Stream. This scenario seems unlikely, because the common periwinkle is a cold-water species.

Others suggested that Norse explorers of around 1000 AD could have brought the common periwinkle as food or bait, explaining their presence in the later Micmac sites. However, a single fossil common periwinkle shell indirectly dated (by dates of other fossils found near it) to 33,000 to 44,000 years ago was reported from a site not associated with humans in southwestern Nova Scotia. At first it appeared that this specimen could not have descended from a Norse introduction, but recent consideration of the archeological methods used to uncover it shows that it could well have fallen into older deposits from much younger ones, including those of the Norse era. No other fossil shells of the species have been found despite thorough exploration of North America, even though many fossils of the species are found in western and northern Europe. Shells of the common periwinkle are also very common in European archeological sites. Yet another hypothesis that would explain both the existence and dearth of the most ancient records and the sudden, explosive spread of the species in the 19th century is that the common periwinkle was present in North America, either native or as a truly ancient introduction, and then went extinct and that the species recolonized after transport from Europe in the 19th century.

Molecular research to tease apart these explanations began in 1977, first showing a severe genetic bottleneck (Chapter 5.1), probably in the distant past. A more recent study examining DNA sequences concluded that the snail would have to have arrived in North America about 8,000 years ago to explain the observed amount of sequence divergence from European specimens. However, critics of that study show that not nearly enough European specimens had been sequenced to draw such a deduction and that, in fact, the observed genetic data are most consistent with an introduction between 200 and 1,000 years ago. This conclusion is buttressed by recent molecular research on both the periwinkle and its most common parasite in both Europe and North American, a trematode (fluke) known as black spot. For both species, genetic diversity was far lower in North America than in Europe, and for neither species was the American population genetically distinct from the European one. Another recent strong hint that the

American population is introduced comes from a comparison of parasitic trematode diversity between the common periwinkle in North America and two native relatives, the rough periwinkle and the flat periwinkle. The common periwinkle had significantly fewer parasites. Had the common periwinkle established in North America around the same time as the two native snails, one would expect all three species to have about the same number of parasites. A final recent piece of evidence in favor of the introduced status of the North American population is that, although the North American and European populations are quite similar genetically, the situation in the rough periwinkle is just the opposite, with substantial differences between North American and European populations.

For this species, then, historical, archeological, ecological, and genetic evidence was marshaled and debated for years before a convincing case for introduced status could be made. Fortunately, for most terrestrial species historical and museum records alone suffice. For marine species, the situation is not as good, and many species remain cryptogenic (Chapter 1.1). For Europe, many introductions occurred well before 500 years ago (Chapter 1.1), but evidence is available for most species to say whether they are native or introduced in particular places.

Some species do not fit comfortably in the categories *native* and *introduced*, not because of absence of historical or other data as in the case of the common periwinkle but because of particular aspects of their history, such as reintroduction in a site after local extirpation. The histories of the capercaillie in Scotland and mountain wisent in the Caucasus (Chapter 1.1) exemplify the problem; both species died out in that part of their native range but were reintroduced from elsewhere within the native range. The mountain wisent case is further beclouded by hybridization among the ancestors of the current herd with American bison. Such populations can be classified only by convention—that is, we can choose to call them native or introduced, so long as we do so consistently. Critics of the attention surrounding invasions often point to such cases as if they cast grave doubt on the entire enterprise of invasion biology and management, but the important point is that such cases are a very small fraction of all populations. For the great majority of species in the great majority of places, either they are native or they got there with human assistance (deliberate or inadvertent), so they are introduced. It may take detailed research of the sort described for the common periwinkle to determine the status, but in principle there is an answer, unlike the cases of the capercaillie or mountain wisent, where the designation will be arbitrary.

Several philosophers have suggested that, under certain circumstances, a species known to be introduced should be accorded an honorary native status, though they disagree about the criteria. For Ned Hettinger, a nonnative species would have to coadapt with the native plants and animals and adapt to the physical environment while maintaining itself without human help. John Rodman requires more—interdependence with the native species and some sort of control by native species. Hettinger would not allow an honorary native to harm a native species, whereas Rodman would impose no such constraint. As will be seen, ecologist Mark Gardener would confer native status on introduced plants that simply cannot be reduced or eradicated, even if they cause enormous ecological damage. Although invasion biologists would question the concept of honorary native, they would certainly view introduced species that are long established (such as the European archeophytes discussed in Chapter 1.1) as less worthy of concern and possible management than more recent invaders. For one thing, invaders that are long established probably did whatever damage they are going to do long ago, and there will not be surprises of the sort discussed in Chapter 4.1. Thus, whereas a new invader might warrant a preemptive attack to lessen a high risk of damage, a long-standing invader would not. Also, some long-standing introduced species have been so common for so long that few people realize they are nonnative. Some, like Kentucky bluegrass (native to Europe, Asia, and North Africa), become cultural icons.

11.4 Are actions against introduced species xenophobic?

Invasion biologists and managers have been taken to task almost since the inception of the field in the late 1980s on the grounds that antipathy to introduced species smacks of xenophobia or racism. This charge takes two forms. The stronger form, leveled by social scientists and some landscape architects and historians, is that activities against nonnative species actually do reflect underlying xenophobia on the part of those carrying them out. The weaker form is that, whatever the underlying motives, actions against introduced species and some of the terminology used to describe them can encourage xenophobia toward various groups of people.

Making the stronger claim, historian Philip Pauly drew a parallel between the increasingly restrictive U.S. immigration policies, such as the implementation of national quotas in 1921 and the Immigration Act of 1924, and the earliest American restrictions on importing plants and animals, such as the Lacey Act of 1900 (Chapter 8.2) and the Plant Quarantine Act of 1912. Certainly the early 20th century was a period in which prejudice against foreigners was rampant in the United States, and the immigration laws at least partly reflected that prejudice. However, Pauly presented no evidence for a linkage between the immigration statutes and laws on importing plants and animals, aside from the rough synchrony of the regulations. Nevertheless, he argued that "...attitudes towards foreign pests merged with ethnic prejudices: the gypsy moth and the oriental chestnut blight both took on and contributed to characteristics ascribed to their presumed human compatriots."[1] In fact, the laws dealing with importing plants and animals were drafted in response to specific effects of introduced species on agriculture, forestry, and nature; impacts of the chestnut blight (Chapter 1.3) and the gypsy moth (Chapter 1.3) were devastating and obvious to scientists and the public alike. Other impacts of introduced species that played a role in early American regulation of introductions were those wrought by the small Indian mongoose on birds and mammals in the West Indies and Hawaii (Chapter 2.5) and by white pine blister rust on several pine species (Chapter 5.3).

Two German garden architects, Joachim Wolschke-Bulmahn and Gert Gröning, go still further in criticizing invasion biology. They point to a Nazi campaign during the 1930s and 1940s to "cleanse the German landscape of unharmonious substance"[2] (by which the Nazis meant plant species) as the lineal antecedent to nature gardening and campaigns against invasive introduced species. The terminology used in the Nazi campaign against foreign plants (for instance, by the Office for Vegetation Mapping) was so similar to that used to describe Jews and other targets of Nazi violence that the motivations for both campaigns surely arose from the same source. The charge that modern advocates of native gardening and restrictions on plant introduction are similarly motivated, however, seems as far-fetched as claiming that anyone who wants trains to run on time is inspired by Italian Fascists.

Environmental historian Peter Coates has examined the history of American popular attitudes toward introduced species and found ample grounds for thinking that popular anti-immigrant sentiment in the late 19th and early 20th centuries was sometimes related to objections to introduced species. The house sparrow, introduced in 1853, is an example, inspiring a popular song in 1883, "The Sparrow Must Go." The sparrow was defended in a 1891 poem, "To an English Sparrow," but, as a frequent urban bird, it was generally detested as a feathered version of the Italian and eastern European Jewish immigrants who settled in cities and were targets of xenophobia. However, Coates draws a distinction between science-based concerns with introduced species, based on real risks of ecological or agricultural damage, and popular antipathy, which derived from a willingness to attribute the same despised traits to foreign animals and plants that were perceived in human immigrants.

As for the weaker charge—that negative publicity about nonnative species leads to anti-immigrant sentiment, no matter whether the concern about invasions is warranted on ecological grounds—the evidence is unclear. It would be surprising if some immigrants did not feel uncomfortable about some anti-invasion campaigns, given the florid language often used to describe invaders in news media reports: *foreigners, interlopers, infiltrators, stowaways.* Even the widely used scientific term *alien* can be off-putting, especially during a period of anti-immigrant sentiment and heightened security measures surrounding immigrants. Individuals of Asian descent in Britain are quoted as asking whether *rhodo-bashing* is associated with *Paki-bashing* (in Great Britain, *Rhododendron ponticum* is a damaging invader from southern Europe and southwest Asia). A blog lamenting a project to remove invasive introduced plants in Boca Raton, Florida, was equally graphic: "If we follow that reasoning a little further, we should probably also get rid of the Haitians, the Cubans, the Canadians, and all the people from New Jersey—they are all 'nonnative.'"[3]

Terminology used in popular reports of hybridization between invading and native species can also lead to unease on the part of immigrants. The serious conservation impacts of hybridization by introduced smooth cordgrass, mallards, ruddy ducks, Sika deer, and other invaders (Chapters 3.6 and 5.5) are lamented, and the advent of molecular genetic techniques for recognizing hybrids has led to a rapidly increasing roster of such cases. The negative impacts in some cases cannot be doubted, but when terms such as *genetic pollution, racial impurity,* and *miscegenation* are used to describe the process they automatically brings to mind the cant of racists who deplore interracial marriage.

Sociologist Brendon M. H. Larson has raised another concern about the impact of terminology used in invasion biology. He wishes to eliminate military metaphors from invasion reports: *advancing front, beachhead, bridgehead, war against invasives, battle against species X, targeting species Y,* and the like. The attempt to stop Africanized honeybees from reaching the United States from South America entailed construction of a "Maginot Line" in Central America. Larson even sees the terms *invasion* and *invasional meltdown* as militaristic. The problem he attributes to these military metaphors is twofold. First, because they are imprecise, he asserts that they impede understanding and management of invasions. Second, he believes they are open to the charge of fostering xenophobia (after all, wars are generally against an enemy that is "foreign") and that this charge, in turn, discredits the science.

The first argument is not convincing. All metaphors are to some extent imprecise, yet we are driven to use them in all sciences, not only invasion biology and not only in sciences. Linguists believe that the ongoing evolution of all languages is driven partly by an inevitable use of metaphor. In the particular case of biological invasions, almost all the metaphors that Larson views as militaristic are also used in public health, where we mount a war against cancer, disease X is public enemy number 1, and an advancing front of disease Y must be stopped or slowed, with vaccination and various sanitary improvements as major weapons. Almost certainly the metaphors initially arise from the depiction of the spread of invading species or pathogens on maps, which closely resemble maps of the advance of armies. Even Charles Darwin and Alfred Russel Wallace in the 19th century spoke of spreading introduced species as invading, and newspapers of the period were full of maps showing the movement of armies. In addition to the obvious cartographic similarity, it would be shocking if policymakers and news media did not use militaristic metaphors for many kinds of campaigns. Most causes have enemies, and we have the war on crime, the war on poverty, the war on homelessness, the battle for minds, and government meltdowns. It seems futile to try to curb such writing on the grounds that it is imprecise.

Does such language foster xenophobia? This is difficult to say. For anyone harboring xenophobic sentiments who had already drawn a connection in his or her mind with introduced species, any terminology that heightens concern about biological invasions would likely strengthen the xenophobia. Would a person who either is not xenophobic or sees no relation between human immigrants and introduced species be swayed by martial language to become a committed xenophobe? To answer such a question would require detailed research by psychologists and sociologists; to my knowledge, such research has not yet been performed.

Opposition to introduced species need not be driven by either xenophobia or concern about ecological or other impacts. It may simply be an aesthetic judgment. When botanist Charles Sprague Sargent objected in 1888 to famed landscape architect Frederick Law Olmsted's use of introduced species in plantings along a river in Massachusetts, it was not because of impact, nor is there any evidence to suggest Sargent was xenophobic. His specific objection was that "it is not easy to explain why certain plants look distinctly in place in certain situations and why other plants look as distinctly out of place ... " but that the latter, introduced from other regions, "inevitably produce inharmonious results."[4] Olmsted conceded that "planting far-fetched trees with little discrimination has led to deplorable results." Writing about introduced ornamentals in 1948, conservationist Aldo Leopold described roadside plantings in the U.S. Midwest: "Through processes of plant succession predictable by any botanist, the prairie garden becomes a refuge for quack grass. After the garden is gone, the highway department employs landscapers to dot the quack with elms, and with artistic clumps of Scotch pine, Japanese barberry, and Spiraea. Conservation Committees, en route to some important convention, whiz by and applaud this zeal for roadside beauty."[5] Obviously a critic could complain that these are in fact expressions of xenophobia masquerading as aesthetic judgments. It is noteworthy, however, that both

Sargent and Leopold were extremely astute naturalists with detailed knowledge of plant biology. Leopold, in fact, was an early critic of introduced species on ecological as well as aesthetic grounds (Chapter 1.2), and the two were very closely related in his mind. Leopold is revered in conservation circles for his land ethic, and he simultaneously developed what has been termed a *land aesthetic*, in which aesthetic judgments of landscapes rest not only on surface visual features but also on knowledge of the evolution and workings of ecosystems and their species.

11.5 Are efforts to contain invasions futile?

When Quammen coined the metaphor *planet of weeds*, he did not advocate abandoning the attempt to prevent invasions and to manage or eradicate established invaders. However, the metaphor suggests that it is futile to try to stop invasions, because the forces that cause them—especially international trade and travel—are growing inexorably, and eventually the world will be biologically homogenized and dominated by the same weeds (both plants and animals) everywhere. Following this line of reasoning, ecologist Richard Hobbs and colleagues have recently suggested that it is impossible to attempt to restore many ecosystems to a semblance of their original states and have advocated trying to use novel ecosystems produced by introduced species and climate change-induced range shifts and perhaps engineering these novel ecosystems with further changes to provide specific services desired by humans, such as flood control or agricultural or timber production. Ecologist Mark Davis has similarly argued that the effort to eradicate or even manage many nonnative plant species is a waste of resources because it can never succeed and that, in many cases, it is not evident that the invaders are causing significant harm. Perhaps most striking is a recent proclamation by Mark Gardener, director of the Charles Darwin Research Station in the Galapagos Islands, that much of the long, expensive effort to control nonnative species there is a losing battle and should be abandoned, even as he concedes their ecologically harmful impact. "It's time to embrace the aliens," he says. Pointing to an invader from Asia, he adds, "Blackberries now cover more than 30,000 hectares [74,000 acres] here, and our studies show that island biodiversity is reduced by at least 50% when it's present. But as far as I'm concerned, it's now a Galapagos native, and it's time we accepted it as such."[6]

Is such pessimism warranted? Should we give up the attempt to prevent or manage invasions and reconcile ourselves to a new era, the Homogeocene (Chapter 12)? This is a complex issue, and no categorical response is possible. Several considerations, however, suggest that a sweeping generic surrender is not warranted. Those advocating simply giving up are overly optimistic on two counts and unduly pessimistic on a third.

First, they are too pessimistic about the possibilities of both eradication and reduction of invasive populations. Of course, if there is no way a species can be eradicated, we should not invest in a program to try—unless even a failed effort yields certain benefits. However, many eradication campaigns that would have seemed hopeless just 20 years ago have either succeeded or seem likely to succeed (Chapter 9.3). The eradication of the melon fly in the Ryukyu Archipelago, Caribbean black-striped mussel in Australia, and rinderpest in the entire world are signal triumphs, as is the eradication of rats, pigs, and goats from progressively larger islands. These success stories result from improved technologies plus determined managers, and there is every reason to think that technological advances, both incremental improvements and occasional innovations from totally new directions, will continue.

Another important fact is that many campaigns that fail to eradicate a target pest nevertheless confer substantial benefits by greatly reducing its population. Chapter 10 included many descriptions of successful control of invasive species, at least for the intermediate term and sometimes for much longer, by mechanical, physical, chemical, and biological means. The fact that we cannot, at least in the immediate future, remove every individual of an invasive species does not mean it is not cost-effective at least to try to manage it. It may well be cost-effective to manage an invasive species in perpetuity even without eradicating it, depending on what is at stake and what resources are available. In some instances the availability of free or low-cost labor may render even a very labor-intensive continuing effort worthwhile (Chapter 10). The use of prisoners to control musk thistle is a good example.

The contention that we should not bother to manage established populations of nonnative species that do not appear to be causing damage rests on the unduly optimistic assumption that we can recognize species that are already problematic or are going to become problems. Many introduced species were innocuous for decades before becoming a major problem (Chapter 4.3). In some of these cases, early eradication might well have been feasible. In other instances, damage was occurring from the onset of an invasion, but it was so subtle or different in nature from what we might have expected that it was not recognized until the invasion had spread widely. The phenomenon of invasional meltdown (Chapter 4.2) is also relevant here—an innocuous nonnative like an ornamental fig can be transformed into a damaging invader by a subsequent introduction. All of these common phenomena argue against the casual assumption that an introduced population has minimal impact and is not worth fighting.

Another overly optimistic assumption is occasionally voiced, that, if we give up on removing invasive species and restoring ecosystems to their preinvasion states, we can simply engineer replacement ecosystems to provide the services and products we desire. This is an extremely rosy view of the ability of ecologists and environmental engineers; it encourages an impractical techno-fix solution to a problem that cannot be so easily solved. Simplified ecosystems designed to provide some service, such as agricultural production, inevitably need interventions (for example, pest control) and subsidies (such as fertilizer and irrigation) that make them expensive to maintain and often generate new problems (for instance, pollution or impacts on nontarget species).

Restoration ecologists do not attempt simply to restore a static entity, an ecosystem that resembles one at some arbitrary time in the past and will stay that way forever. Even without human impacts and invasions, ecosystems change—species evolve and systems respond to new environmental conditions. The goal of restoration is to restore an ecosystem to a semblance of an earlier state that will allow it to continue on its own trajectory—to change and adapt without wholesale rapid destruction of its components and without continued human intervention and subsidies. Well-managed ecosystems restored to a semblance of their historical states have often proven cost-effective and served several functions simultaneously (for instance, production of timber and habitat for species of conservation value). The demands of a growing human population already containing 7 billion individuals necessitate large parts of the earth dedicated to engineered ecosystems totally devoted to serving human needs, but this is all the more reason to work hard to restore salvageable ecosystems that have been damaged by invasions and to protect others as best we can from further invasive threats.

Ironically, the Galapagos campaign against damaging nonnative species that Gardener deems a failure is rife with successes. At least 27 invasive species, including several plants, have been eradicated from the Galapagos. Feral goats have been eradicated from many islands, and feral pigs have been eliminated from Santiago Island (Chapter 9.3). The archipelago has also seen successes in long-term management short of total eradication. For instance, the cottony cushion scale has recently been brought under successful biological control by the same Australian insect, the vedalia lady beetle, that controlled the scale a century earlier in California (Chapter 10.3).

Finally, some ongoing management campaigns that require much labor and that will never lead to total eradication of a targeted invader may yet serve an important purpose: inspiring aggressive efforts to deal with invasive species generally and education of the public and its policymakers about biological invasions in general. Many regions have programs sponsored by governments or conservation organizations to mobilize volunteers to remove invasive plants physically—weed pulls, bush bashes, and the like (Chapter 10.1). Such efforts provide a measure of control of harmful invaders, educate many people about impacts of introduced species, and inspire vigilance and confidence that the threat of invasions can be mitigated.

11.6 Should animal rights govern management of invasive species?

Advocates of animal rights have bitterly opposed many campaigns to eradicate or manage introduced animal species. They cite various grounds, including that humans have no right to kill animals or that if they have the right to do so the methods used are inhumane. Even eradication of such a widely hated invader as a rat can be controversial, no matter the evidence of ecological harm. On Anacapa Island in the Channel Islands National Park in California, introduced ship rats were threatening a colony of a bird species, Xantus's murrelet, listed as a Federal Species of Special Concern, as well as an endemic rodent, the Anacapa deer mouse. The Park Service developed a plan to eradicate the rats, taking care first to remove most of the deer mice that would have been susceptible to the same poison baits to be used against the rats. The rugged terrain of the island rendered any other plan infeasible. The Park Service was astounded that publicity about the plan aroused substantial opposition on the grounds of animal rights. Retired Channel Islands National Park superintendent T. J. Setnicka, speaking in retrospect about the controversy, said, "We didn't think we would have much problem in the media with this project. Who could love a rat? As it turned out, lots of people."[7]

A lawsuit by the animal rights group Fund for Animals was eventually dismissed but delayed the eradication. There was even an attempt at sabotage by a quickly formed organization, the Channel Islands Animal Protection Association, which distributed vitamin K pellets as an intended antidote to the poison. The sabotage failed, the eradication succeeded, and populations of both the murrelet and the reintroduced deer mouse have largely recovered.

Arguments based on animal rights are not easily countered, and in the end it may be impossible to reconcile certain advocates with the idea of killing introduced animals even if they are causing demonstrable ecological or other damage. Philosophers approach this issue by asking whether animals other than humans must be given consideration from a moral perspective. If the answer is yes, then the interests of these animals should be considered when we deal with them. Even if individuals of species other than humans are morally considerable, philosophers would not automatically accord them rights, nor would it necessarily mean that individuals of all species are equally considerable either to one another or to humans. Philosopher Clare Palmer notes that two fault lines divide ethicists with respect to moral considerability. One she terms the anthropocentric–nonanthropocentric fault line and the other the individualist–holistic fault line. Although at first blush it might seem that the anthropocentric–nonanthropocentric fault line would separate animal rights advocates from those trying to remove invasive species, in fact most invasion biologists and conservation managers would see themselves not as anthropocentric but as defending nature against harm caused by humans (who are responsible for the introductions); they would therefore probably put themselves on the nonanthropocentric side of the divide.

It is the other fault line that causes the great majority of controversies about introduced species management that are linked to animal rights. The issue is that individuals of a nonnative species are harming those of native species, possibly even threatening extinction of the latter. Many holists would see a species as morally considerable above and beyond whatever consideration is due to individuals of that species. A philosophical individualist, by contrast, would not grant moral considerability to a collective, only to individuals. Thus, a holist would allow killing individual rats to prevent danger to a bird species. Probably most conservationists taking this view would desire or even demand the most humane sort of death possible for the rats, but primacy would go to saving the species the rats threaten, as the rat species is widespread with many populations and as eradication of one rat population on one island in no way threatens the rat species. Individualists, on the other hand, would reject the campaign, to an extent determined by the exact criteria used to determine moral considerability and how one assesses relative degrees of considerability. Almost all animal rights advocates grant moral considerability to all mammals, and most would grant it to birds as well. Often sentience—the ability to sense or feel—is seen as the key criterion. Philosopher Peter Singer, for instance, believes that an animal that can suffer is owed moral considerability; probably this ability would extend to all vertebrates, not just mammals and birds. He also believes individuals of all sentient, self-conscious species are equally considerable. Others see sentience as the key criterion but are not as egalitarian as Singer. Still other philosophers, such as Nicholas Agar, allow that both species and individuals have moral considerability but that species are more considerable.

Whatever one's assessment of degrees of moral considerability, for it to be implemented in some sort of action it must be associated with what philosophers call a moral theory, and it is certain moral theories that confer rights. Many moral theories are termed *consequentialist* because they state that actions should be based on expected consequences that maximize the ratio of good consequences to bad consequences, averaged over all animals that are morally considerable. Other moral theories confer rights that cannot be overridden by any maximization calculation, and one such right for morally considerable individuals, in the eyes of many rights theorists, is the right not to be killed. In this view, no matter how many individual birds or even species of birds could be saved by eradicating a rat population, it is morally improper to do so because the individual rats have a right to live. Many rights theorists would additionally point to the fact that rats are not moral agents and humans are. That is, rats cannot be held morally accountable for killing birds, but humans are morally accountable for killing rats. Further, the rats are not even responsible for being in most of the locations where they are pests; they got there because of human activity and in some instances (e.g., Pacific rats introduced by the Lapita people to Pacific islands) may have been deliberately introduced. In general, these arguments can be marshalled in opposition to any campaign against an introduced species that is an eradication or management target. They were introduced directly or indirectly by humans, and they are not morally culpable for doing what they do naturally to survive and reproduce.

Methods used in eradication projects, and even in some types of maintenance management, may also cause controversy even among holists who do not grant individual introduced animals a right to live that trumps continued existence of a native species. For example, methods that cause a painful death, such as snares for pigs or chemical baits such as 1080 or brodifacoum (Chapters 10.1 and 10.2), not only infuriate individualists, particularly those for whom sentience is a key criterion, but also upset holists who simply do not want to see animals suffer. Some animal rights advocates call for chemosterilants as a more humane way to rid a site of invasive vertebrates, but, even aside from the fact that for most species such chemicals do not exist, their use would not satisfy everyone. Philosopher Nicholas Agar, for example, sees reproduction as a fundamental right of animals.

A long-standing controversy in the Hawaiian Islands epitomizes animal-rights based objections to eradication and even maintenance management. Pigs were introduced twice to the archipelago, initially at least 1,200 years ago by the Polynesians who first colonized Hawaii. These pigs were a small variety descended from Asian pigs and were widely introduced in the Pacific by Polynesians for food. In 1778, European boar were introduced by Captain Cook, and many varieties have been introduced since then as game animals. The two types are interfertile, have interbred, and have wrought ecological havoc. Their rooting erodes the mountainsides of the archipelago, and the pools the rooting creates aid reproduction of introduced mosquitos that vector avian malaria, which is a main threat to the existence of several native bird species (Chapter 3.5). The rooting also aids invasion by nonnative plants, especially firebush and strawberry guava, that transform whole ecosystems and may ultimately threaten the existence of some native species.

Managers of the U.S. National Park Service as well as the Nature Conservancy land stewards responsible for maintaining large natural areas of Hawaii have mounted campaigns to control and possibly eradicate pigs from large areas. The only feasible method in the mountainous terrain is snaring, a painful method that especially angers animal rights advocates. In addition, the Nature Conservancy and federal and state agencies have fenced areas cleared of pigs to keep them from reinvading, which has infuriated hunters, especially native Hawaiians whose ancestors have hunted pigs for centuries. Oddly, the animal rights advocates, led by the nongovernmental organization People for the Ethical Treatment of Animals, have allied with hunters, normally their opponents, to try to stop pig eradication in Hawaii. (A similar unusual alliance was established between the Fund for Animals, supporting animal rights, and hunters trying to stop a project by the National Park Service to eliminate nonnative mountain goats from Olympic National Park in Washington State.)

Another controversy based on animal rights erupted in Hawaii over a plan to control an introduced Puerto Rican tree frog, the coqui. The dense populations of coqui are believed to threaten native insect populations, and in addition the frogs are viewed as a pest because of their loud, bell-like nocturnal calls. However, animal rights advocates hindered control operations and are suspected of having introduced the frogs illegally to new areas. In New Zealand, animal rights advocates have opposed efforts to control the brushtail possum, an introduction from Australia that is New Zealand's most damaging invader, affecting both conservation interests and agriculture. They browse certain native tree species, changing forest composition and physical structure. They also compete for food with native birds and eat their eggs. As vectors of bovine tuberculosis, they are a pest to dairy farming and ranching. The controversy is particularly heated because the chemical bait 1080 is the only effective method of controlling their populations in dense and mountainous forest habitats. In San Francisco a campaign to remove a flock of cherry-headed conures (a South American parrot species) that may harm native birds was stymied by opposition from animal rights advocates on the grounds that conures are intelligent and sentient animals. Similar objections have hindered attempts to eradicate monk parakeets in Florida, California, and other states, even though their nests cause expensive damage to electrical utility structures.

Campaigns to eradicate some plant species have also generated enormous controversy. Usually this is because they are perceived as aesthetically pleasing or have acquired cultural significance through long presence at a site. However, some philosophers, such as Robin Attfield, view plants as morally considerable, and occasionally removal of introduced plants is opposed at least partly on grounds of plant rights to live. In several cases in California, removal of eucalyptus trees, which have been present since the 19th century, has generated objections remarkably similar to those concerning introduced vertebrates. For instance, in 1979 the U.S. National Park Service proposed to remove most of the eucalyptus trees from

Angel Island in the Golden Gate National Recreation Area. Eucalyptus trees harm several native plant and animal species in California, especially by propagating fires. But they have become a cultural icon—a style of landscape painting popular in southern California between 1915 and 1930 is even known as the eucalyptus school of art. The Angel Island project was delayed for five years by members of the organization Preserve Our Eucalyptus Trees, who attacked what they termed *plant racism* and called Park Service employees plant Nazis. Many of the protesters felt that the large trees simply deserved not to die. Similarly, a woman protesting a eucalyptus removal project in southern California argued that it was immoral. Although her husband undercut this plant rights arguments with the statement, "I think this is not about trees, it's about views,"[8] concerns over aesthetics and property values need not exclude moral revulsion.

Explicit plant rights statements are more frequently attributed to tree sitters trying to protect native redwoods and Douglas fir trees in the Pacific Northwest than to defenders of introduced plants targeted for management or eradication, but the same ideas do surface in some noteworthy cases. For example, a major controversy concerned a proposal in 1996 to restore 7,000 acres of woodlands and shrublands in suburban Chicago to the oak savanna and prairie that had dominated the site before European settlement. Much of this area has been invaded by common buckthorn and glossy buckthorn from Europe, but many people opposed the project. Most objections were on the grounds that the trees were an amenity, providing shade and beauty. However, moral objections were also raised, as by a group opposed to the restoration on the grounds that "God made those non-native plants and trees, just as surely as he made oak and trillium."

12

PROSPECT—THE HOMOGEOCENE?

12.1 What is biotic homogenization?

Multitudes of species have been introduced to all parts of the earth (even Antarctica has about 10 established invaders) in what ecologist Francis Putz has branded the Homogeocene. The main change marking the end of the Holocene (the current epoch) is the geographical homogenization of the global flora and fauna, almost all caused by one species, *Homo sapiens*. Invasion biologists have characterized this biotic homogenization by comparing numbers of species shared between sites today with the shared numbers before humans started moving species around. For instance, Frank Rahel has studied fish faunas of states in the continental United States. On average, pairs of states now have 15.4 more freshwater fish species in common than the same areas did before European settlement. Previously, 89 pairs of states had no fish species in common; they now share on average 25.2 species. David Quammen sees the endpoint of this homogenization as a planet of weeds, in which the same weedy plants and animals will characterize all sites (Chapter 11.2). Are future generations indeed fated to live in a biotically homogenized world?

Many factors suggest that this prediction may not be far off the mark, including human activities that transport species to new locations and other human activities, acting over the longer term, that will make it likely that a greater fraction of these arrivals will establish populations and spread. Acting against these factors are increasing public recognition of the ecological and economic toll taken by invasions and ever improved technologies to prevent, eradicate, or mitigate invasions. Which set of factors will prevail? These forces will be discussed in the following sections.

12.2 How will patterns of human activity affect biological invasions?

The forces that carry species to new places—trade and travel—are growing rapidly. Shipments of cargo are tallied in units of ton-miles—that is, 1 ton of cargo moved 1 mile. From 1970 through 2010, the amount of cargo shipped by sea approximately tripled, from less than 12 trillion ton-miles per year to more than 33 trillion ton-miles per year. If this rate of growth were to continue for the next 20 years, the amount of seaborne cargo would be about 57 trillion ton-miles in 2030. Cargo shipped by air is predicted to increase even more, by about 6% annually for the next 20 years, from 110 billion ton-miles in 2010 to 355 billion ton-miles in 2030. Passenger air travel has increased on average by about 5% annually for the last 30 years in terms of number of passenger-trips (one passenger taking one trip), and it is expected to increase at between 4% and 5% for the next 15 years, which would approximately double the current annual figure of about 2 billion passenger-trips. Of course, numbers of cargo shipments and passenger-trips fluctuate greatly from year to year depending on economic conditions. For instance, all these statistics plummeted during the economic crisis of 2009 and then began recovering by 2010 or 2011. However, the long-term averages have been constantly upward, and predictions are that they will continue to increase at roughly the same rates for at least the next one or two decades.

These are sobering data. If no changes occur in how cargo and passengers are monitored and regulated for movement of living organisms, we might expect a concomitant increase in the number of introductions and, ultimately, the number of damaging invasions. In fact, there are reasons to think that the situation will be even worse. Species are often being transported with greater numbers of individuals than would have been possible even 50 years ago, enhancing the probability that a population will at least establish initially. Perhaps most significantly, an increasing number of populations of invasive species have been founded by mixtures of individuals from different regions, with different genetic backgrounds and different evolved adaptations. Chapter 5 gave the examples of the multicolored lady beetle, reed canary grass, and the Malaysian trumpet snail as species for which genetic recombination among individuals introduced from different regions appears to have produced more invasive types of offspring.

It is widely recognized that global warming would likely continue for decades even if emissions of carbon dioxide and other greenhouse gases were suddenly subjected to the most stringent policies proposed so far. A similar but less widely known "overshoot" would occur with biological invasions even if we were suddenly to enforce extremely strict regulations to prevent introductions. Some established introduced species that are not presently invasive will surely become invasive if past experience is any guide. Many invasions occur only after a lag that may extend for decades (Chapter 4.3). Sometimes the lag ends because of physical environmental changes, other times because of introduction of other species (the meltdown phenomenon), and other times for reasons we do not yet understand.

Franz Essl and colleagues have hypothesized the existence of an invasion debt. Observing that current numbers of invasive species are statistically more closely correlated with indicators of socioeconomic activity in 1900 than with the same indicators in 2000, they suggest that this means invasions of past decades have not yet produced their full impacts. For instance, plants that have become invasive in Europe, even if they started spreading quickly and did not undergo a period that might be termed a lag, have taken well over a century on average to reach boundaries at which their spread greatly slowed down or stopped. Extinction of some native species will be just one future impact of this invasion delay. Chapters 3 and 4 describe many other impacts on species, ecosystems, and human society.

12.3 How will global climate change and other global changes affect biological invasions?

Four great global changes are occurring today: biological invasions; climate change; enhanced geochemical cycles (especially the carbon and nitrogen cycles); and land use change (particularly the transformation of many natural habitats to residential, commercial, and other anthropogenic habitats). However, these factors do not act alone—they can influence one another and generate impacts that are more than additive. How will changed climate, geochemical cycles, and habitats affect biological invasions?

Rising temperatures and other predicted climatic changes (for instance, increasingly severe and frequent droughts in some places) will surely change the ranges of many established invaders and allow other introductions that would not previously have survived and spread to do so. Various statistical methods lumped under the rubric of species distribution models (SDMs) can be used to predict which species will spread where under various climate change scenarios (Chapter 7.2). One such study in North America by geographer Bethany Bradley and colleagues has already predicted that three prominent invaders—kudzu, privet, and cogongrass—will all spread greatly. Similarly, species currently restricted to warmer parts of the United States, such as the red imported fire ant, are expected to expand their ranges. Of course other established introduced species, such as those adapted to cold climates, might see their ranges retract. And some species that might have established populations in the current climate will not do so when it is warmer, or in areas where it becomes drier. However, researchers who have studied this issue generally agree that more invasions will be fostered by the changing climate than will be hindered by it.

An invasive pathogen of ruminants (especially sheep and some deer species) that has recently spread much farther north in Europe provides a foretaste of the sorts of changes that may be in store. Until recently, bluetongue virus was restricted primarily to Africa and south Asia and was not established in Europe, although sporadic outbreaks occurred in the southernmost regions. In Africa, the main reservoir of the virus is cattle, and it is transmitted among cattle and also to other species by a midge (a tiny fly). Until 1997, this midge was found in Europe only in Portugal, southwestern Spain, and some Greek Islands. However, since then, outbreaks have occurred in 12 different European nations, some 500 miles farther north than the disease had ever been seen before. A team led by ecological modeler Bethan Purse attributes the spread of bluetongue to climate change, in particular the northward expansion into Europe of the midge and increased virus persistence in the winter. Now that the virus is in Europe, it is also transmitted by new flies, European relatives of the original vector. In short, warming aided the virus itself and helped spread the original vector and in addition brought the virus within range of new vectors. There is every reason to think that other vector-borne pathogens may similarly spread as climate becomes more suitable for them or for their vectors.

Changes in nutrient cycles are also likely to favor certain invaders. The change in the carbon cycle as atmospheric carbon dioxide increases is the main driver of global climate change, which, as has just been described, will affect many invaders. In addition to this indirect effect, changes in some nutrient cycles will surely aid some invaders to establish and spread. In Chapter 3.1, the role of the nitrogen-fixing invader firebush in Hawaii was described, particularly its aid to other introduced plants that had been limited by low nitrogen in the nutrient-poor volcanic soil. An experiment in Minnesota grasslands has similar implications; nitrogen-fertilized plots were quickly overgrown by European quackgrass, and many native grass species disappeared. The earth is currently undergoing a great intensification of the nitrogen cycle, from both aerial sources (for instance, burning of fossil fuels) and terrestrial ones (fertilizer and livestock and poultry wastes). Parts of North America, Europe, and Asia have staggeringly high nitrogen deposition rates, greater than 67 pounds per acre per year, more than 100 times the natural rate. In addition to causing a number of native plant species to expand their ranges, this deposition enhances invasion by nonnatives. For instance, in the United States, giant reed preferentially invades sites with greater rates of nitrogen deposition.

Human-disturbed habitats as well as novel sorts of habitats (such as agricultural fields or suburban yards) are typified by high numbers of introduced species and reduced numbers of natives, because disturbance often liberates resources that invasive species are well adapted to garner and because native species in a site are no better suited to a novel habitat than are newcomers (Chapter 2.6). Thus, the increasing conversion of natural habitats to human-dominated ones is highly likely to increase dominance by invasive species at the same time as it threatens native ones. The increasing spread of human-dominated landscapes is combined with the fragmentation of natural habitats such as forests or prairies, which are increasingly restricted to habitat islands surrounded by agricultural, residential, and other novel habitats. The flow of seeds and individual animals from the surrounding human-dominated landscape will be a constant threat to the native community. Simply setting aside small habitat fragments as refuges or living museums as a way to preserve native species and ecosystems, as suggested by philosopher and invasion biology critic Mark Sagoff, is an impractical approach. Decades of experience by refuge managers have demonstrated that the ecosystems do not remain intact and that, particularly for small refuges, intensive ongoing management activities (prominent among them the removal of invasive introduced species) are required to conserve a semblance of the original ecosystem. Invasion into reserves dominated by closed canopy forests is delayed relative to invasion into grasslands, because the main opportunity for plant invasion is when a canopy tree dies, leaving a gap that various seeds can colonize. Because many canopy tree species are very long-lived, these opportunities arise infrequently. But even in such habitats, invasion is a constant threat to isolated fragments, and vigilance and active management are required. For instance, in many closed forests of the northeastern United States, Old World Norway maple and tree-of-heaven invade and become canopy trees, while Japanese stiltgrass and Japanese knotweed tend to dominate the ground cover if they become established.

12.4 What new technologies will aid detection and monitoring of invasions?

Remote-sensing methods, using aircraft or satellites, have been deployed since the mid-1990s to track plant invasions. The trick is to have sensors with fine enough spatial resolution and also spectral (color) resolution to be able to pick out non-native plants surrounded by native vegetation. Early efforts were usually inadequate on these counts, restricted to working well only if the timing of some nonnative plant differed from that of natives. For instance, Landsat satellite images were used to map Amur honeysuckle, an Asian shrub, in deciduous forests in Ohio in late autumn once native trees had lost their leaves. However, in most instances, neither the spatial nor the spectral resolution sufficed for such purposes.

High-resolution sensors are now changing this situation. For instance, weed scientist James H. Everitt and his colleagues found that the flower colors of Eurasian leafy spurge and Asian salt cedar were distinct enough to map infestations in the American West using high-resolution sensors. More recently, imaging spectrometers (also known as hyperspectral imagers) gather information on the full spectrum across much of the visible and shortwave regions, and satellites with these instruments have made this approach the main one for studying plant invasions. In this method, detailed spectral profiles are gathered for both native and nonnative plants, and this information allows researchers to focus on the spectral regions that best distinguish the target invasive species. Hyperspectral imaging can even be used to study chemical changes and nutrient fluxes in invaded ecosystems. For example, ecologists Gregory Asner and Peter Vitousek used this method in Hawaii to show how introduced firebush and kahili ginger changed nitrogen and water concentrations in tree canopies. Hyperspectral imagery is now also being combined with LiDAR, the same remote-sensing technology used to detect speeding cars. LiDAR measures distance to a target very accurately, so it can determine physical structure, while hyperspectral imagery determines color. Perhaps the greatest challenge facing users of the most modern remote-sensing techniques is not limitation on the resolution of the data that can be gathered but the sheer amount of information provided by advanced sensors. Often complicated data-mining computer algorithms are required to detect the signals of invasive plants in the welter of gathered data.

Dramatically improved methods of sequencing nucleic acids also open new windows to detecting and monitoring invasions. The Human Genome Project, initiated in 1990, took 13 years and cost \$3 billion to sequence the human genome. Nowadays, with new technologies lumped under the term *next-generation sequencing*, the same task could be accomplished in a couple of hours for perhaps \$100. These new techniques have already been applied to invasions in two main ways. First, they have been the chief means by which many cryptic invasions (Chapter 2.2) have been discovered. This effort has been aided by the Canadian Centre for DNA Barcoding, a project initiated in 2003 that aims to catalog all life by sequencing a particular gene in all species and recording these sequences as barcodes. Thus, DNA from an individual of an unrecognized species can be sequenced, the sequence can be compared with the barcode, and one can tell if the individual belongs to a known species (and, of course, whether that species is introduced in the area where the individual was found). The same data can be gathered from several individuals to see if they all belong to one species or to several. It is now evident that, although the particular gene used by the Canadian Centre suffices for very many species, sequences of other genes will have to be used in some instances. However, the principle still stands: rapid sequencing of a bit of tissue of an individual to find out quickly whether it belongs to an introduced species.

Nucleic acid sequences can be used not only to determine if collected specimens belong to native or nonnative species but also to determine if a particular species is even present. Environmental DNA (known as eDNA) is DNA that came from a living organism but is now simply in cells shed by that organism into the environment. Dead skin and waste products, for instance, contains DNA. Nowadays, for example, wildlife researchers can determine which species left a fecal pellet by finding the DNA in cells shed by the animal with the feces. To use eDNA to detect invasion of a target species, one first employs primers, short strands of DNA that pair off (hybridize) with the complementary strand of the DNA of the target species. A molecular method known as polymerase chain reaction (PCR) uses these primers to produce vast amounts of the target DNA. The method is highly sensitive— the primer will find even minute amounts of the target DNA, and the PCR process then multiplies this amount manifold so the researcher can detect its presence. This approach promises great improvements in early detection of invasive species, so long as one is on the lookout for a particular invasive species and not just scanning the environment to see if any invasive species is present. A recent use of the technique in Europe detected American bullfrogs when the latter were present at such low densities that visual sightings and detections by their calls failed to find any. In principle, this method could be very widely used, with water samples routinely taken and tested with primers for a number of possible invasive species.

12.5 How will new management technologies affect invasions?

Many invaders have been completely eradicated, and the scope and complexity of successful eradication projects have increased dramatically, with rats eradicated from islands of over 27,000 acres and goats and pigs eradicated from islands of over 120,000 acres (Chapter 9). In addition, an increasing number of invasive species, even though not eradicated, have been brought under good long-term control by physical, mechanical, chemical, and biological means (Chapter 10). In many cases, these are invasions that would have seemed intractable as recently as a decade ago. The majority of these eradication and maintenance management success stories have resulted from the incremental improvement of technologies that have existed for a long time—improved traps, more effective toxic baits with fewer nontarget impacts, and the like. In some cases, an improvement was more than incremental and greatly enlarged the scope of possible uses. For instance, the addition of a hormone injection to keep Judas goats in estrus (Chapter 9.3) was a major advance and increases the size of islands that can be cleared of goats (and possibly other mammals) several-fold.

Other techniques are totally new; they often come from research in fields not previously directly related to biological invasions. For example, fish behaviorist and endocrinologist Peter Sorensen and his colleagues have recently developed a completely novel method to manage sea lampreys in the Great Lakes region (Figure 12.1). Knowing that sea lampreys are anadromous (adults return from the ocean or large lakes to streams breed) and observing that they rarely enter streams with no larval sea lampreys, Sorensen deduced that the larvae must emit a chemical called a pheromone that attracts adults. His team first concentrated the pheromone in water the larvae had been in and determined that he was correct—it attracted adult sea lampreys. Next they determined the chemical structure of the pheromone and synthesized it. A minute amount attracts lampreys from a great distance, and they can then be lured into traps. This method is increasingly used in lakes of the upper Midwest with great success. Another high-tech approach from a radically new direction was pioneered by aquatic ecologist David Aldridge, who observed that zebra mussels avoid being poisoned by potassium chloride because they close their shells when they detect even minute amounts of this chemical and thus avoid taking in toxic doses. Aldridge

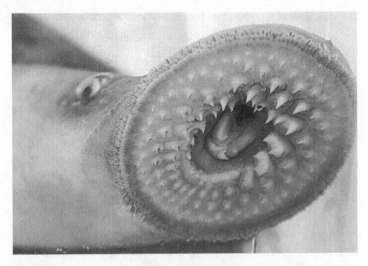

Figure 12.1 Mouth of sea lamprey, introduced to Great Lakes. (Photograph courtesy Peter Sorensen.)

made microscopic beads packed with potassium chloride and coated with a fatty substance. When released in water facilities clogged by zebra mussels, these beads do not elicit shell closure by the mussels, because the fatty coating masks the potassium chloride. The beads are taken in by the mussels, which gather their food by filtering massive amounts of water. The beads rupture and the potassium chloride kills the mussel without nontarget effects, as the minute amounts of the salt are diluted to harmless concentrations in the water.

Not all effective novel approaches are high-tech. In Hawaii, a barge-mounted vacuum device known as a Super Sucker has cleared coral reefs of two introduced fouling red algae, gorillo ogo and spinosum. These had been deliberately introduced in the 1970s to produce carageenan, an emulsifier and fat substitute, but had become highly invasive. Hand clearance was found to be prohibitively expensive, but the Super Sucker operated by two divers for one hour removes 800 pounds of algae, an amount that previously would have taken 150 volunteers and 10 divers working for the same length of time. In the early 2000s, the Scottish agency responsible for controlling introduced American mink developed a remarkably effective low-tech breakthrough called a mink raft. This consists of a wooden floating platform with a thin layer of clay that retains tracks of mink that climb over it. The platform is anchored and monitored to determine the presence and density of mink in an area, which allows much more effective placement of traps than had previously been possible. Mink rafts are easily constructed but are now even available commercially, and their use has permitted the Scottish campaign to lower the range and density of mink greatly in certain areas.

A number of genetic approaches to invasion management are currently under development. During the late 1960s and 1970s, several researchers suggested a totally different way to control pest insect species (for the most part, the targets were introduced species) called autocidal control. The underlying idea was to harness the forces of evolutionary genetics so that the target species would sow the seeds of its own destruction without interactions with predators, pathogens, or parasites and without the use of chemicals or manual labor. This approach was probably inspired by Knipling's successful eradication of screwworms from the island of Curaçao in 1954 by the sterile male technique (Chapter 9.3). The fact that the fly's own mating behavior could be turned against it galvanized entomologists and geneticists. One possibility suggested at the time was to use an evolutionary process known as *meiotic drive* to attack pests and perhaps even drive them to extinction. In sexually reproducing organisms, both the male and female parents, which are diploid (that is, have two copies of each chromosome), have cells that produce gametes (sperm and eggs) that are haploid (have just one copy of each chromosome) by a process called meiosis. Normally, which of the two copies of a chromosome gets included in the gamete is a matter of chance, and each has a 50% probability of being included. However, it was observed that in certain instances one particular chromosome was much more likely to be chosen than the other (presumably because of some gene it was carrying). That chromosome was said to be meiotically driven.

This observation led to the idea that, if somehow a driven chromosome could carry a gene that would harm the pest population, the population would automatically be reduced. Of course, as soon as the detrimental gene did whatever it was going to do, it would be "seen" by the environment, and natural selection would operate against the individual carrying the gene, so the gene would disappear from the population. Suppose, however, that the gene could not to be seen except under certain circumstances, after it had spread throughout the population? An example might be a gene making a pest insect susceptible to some insecticide, but the insecticide is not normally present in the environment. Such genes are known as *conditional lethals* because they work only under certain conditions, and at that time they become lethal to their bearers.

The idea would be to use meiotic drive to spread a gene for insecticide susceptibility and then, when the whole population had the gene, to spray the insecticide.

However, this idea faded as research on meiotically driven chromosomes always seemed to show either that they were associated with a selective disadvantage or else that natural selection quickly selected for genes on other chromosomes that counteracted whatever was causing the driven chromosome to be driven—that is, to return the process of meiosis to an equal chance for each chromosome to be included in the gamete. Another drawback turned out to be that it was very hard at the time to find conditional lethal genes in the target species of interest, which were mostly flies (including mosquitoes). Now, 50 years later, advances in genetic manipulation have led to a resurrection of this idea.

A related idea was to distort the sex ratio of a target pest using meiotic drive. In most species, in males, one chromosome pair consists of two different sex-determining chromosomes, one of which (the Y chromosome) causes the bearer to become a male. Females have two copies of the other chromosome, the X chromosome. In humans, for instance, if the sperm carries a Y chromosome, the fertilized egg becomes a boy; if it carries an X chromosome, the fertilized egg becomes a girl. (In birds and a few other species, it is the female that has two different sex chromosomes, while the male has two of the same [male-determining] sex chromosomes). Normally, at meiosis, 50% of the sperm (in birds, 50% of the eggs) receive the male-determining chromosome (in birds, the female-determining chromosome), and 50% receive the female-determining chromosome. However, instances were found in which one or the other type of chromosome was driven at meiosis—that is, for instance, the male-determining chromosome was found in many more than half the gametes. The idea, then, would be to introduce such a driven sex chromosome into a pest population, and, over time, the entire population would be male (or female), leading to its demise. This result would not be unlike that of Knipling's successful sterile-male technique, in which females eventually find no fertile males with whom to mate.

However, once again, natural selection seemed to defeat this scheme, as the presence of a driven sex-ratio distortion gene seemed always to be countered by natural selection for genes on other chromosomes that neutralized its effects and returned the sex ratio to 50–50. Or else, perhaps such a gene actually caused the population carrying it to go extinct so quickly that scientists never got to see it.

After about a decade of failures with such autocidal ideas, most pest insect researchers lost interest in them (except for the sterile male technique) and turned primarily toward chemical and biological control. However, today, in an era of transgenes (genes from one species inserted into the genome of another species) and many advances in genetics and cell biology, similar autocidal ideas are again attracting attention, especially for control or eradication of invasive insects (such as crop pests and mosquitoes that vector diseases) and invasive fish (which, as was discussed in Chapter 9.3, have proven especially difficult to manage).

Biologists Richard D. Howard and William M. Muir in 1999 proposed a transgenic approach called the Trojan gene method. The idea would be to use a transgene, or a tightly linked cluster of them, that has two effects—it would increase male mating success while lowering the fitness of the offspring of such a male. The fact that one can now use transgenes from many species instead of having to isolate genes within the target species greatly increases the possibility that such a gene might be found, and in principle such a gene could extinguish a target population very quickly. However, as Muir and Howard pointed out, natural selection could act quickly to counteract such a Trojan gene in two ways: selection of modifiers to act against the viability reduction of the offspring; and selection in favor of females who would discern which males were transgenic and which were not and would mate with the latter. In short, the project would have to succeed very quickly in eradicating the pest, or natural selection would defeat it. In a way, it would be a race between geneticists isolating and moving transgenes and pest species evolving immunity to their effects.

Another approach entails manipulating entire sets of chromosomes. For instance, triploids (that is, individuals with an extra set of chromosomes) have been produced for over 20 years in grass carp and the Pacific oyster, among invasive species. Triploids generally have greatly reduced fertility or are even sterile, and for oysters triploids have sometimes been planted with the goal of preventing cultivated individuals from establishing self-sustaining populations. Unfortunately, this is not a foolproof method for oysters because a few triploid oysters revert to diploidy, triploid female oysters produce some fertilizable eggs, and screening to eliminate diploid oysters before establishing the cultivated populations is not always perfect. With grass carp, which are unselective and browse native plants as well as introduced ones, the method is more promising. Triploid grass carp can be produced by heat, cold, or pressure shock to fertilized eggs, and their triploid status is easily checked by automated examination of their blood cells, which have larger nuclei than those of diploid carp. They have often been used with some success in an attempt to garner the temporary benefit of clearing a weed-infested area without the risk of establishing an invasive population. Although a few diploids have occasionally been accidentally released, this has not been a large problem.

A method recently suggested by biologists Juan B. Martinez and John L. Teem has been termed the Trojan chromosome approach. Fishes have an XY sex determination system, just as humans do, but it has been observed that in some species, such as chinook salmon and Nile tilapia, some individuals with Y chromosomes are actually females, presumably because of exposure to female sex hormones during development. In fact, in commercial aquaculture, ordinary male tilapia are feminized by addition of diethylstilbesterol (a synthetic estrogen) to their feed, resulting in females that nevertheless have the male chromosome complement of X and Y. In aquaculture, these feminized XYs are then mated with normal males, and some offspring are supermales, with two Y chromosomes. These supermales are then mated to normal XX females so all male progeny (XY) will be produced, which is desirable in aquaculture because the males grow faster. However, these same YY supermales can also be feminized by exposure to female sex hormones, producing females that are nevertheless YY. What Gutierrez and Teem noticed was that, since these females could produce only Y-bearing eggs, all offspring of a mating of such a sex-reversed supermale (now a female) with a normal male would be males, either XY or YY. Over time, the population would become increasingly male, and the process could be sped up by repeated stocking of such sex-reversed YY supermales. Eventually, when only males were left, the population would disappear! The method has not been tried in practice yet, but it seems technologically feasible. The rub would be that the sex-reversed individuals might not have the same likelihood of surviving or mating as normal individuals.

A project under way in Australia headed by ecologists Nicholas J. Bax and Ronald E. Thresher involves using daughterless carp to reduce or extirpate common carp populations. A transgene is introduced into individuals and disrupts a step in sexual development (expression of the aromatase enzyme) so that all offspring are male. The heritable transgene increasingly skews the sex ratio toward males, and eventually reproduction declines for lack of females; ultimately extinction would ensue, at least if natural selection did not somehow intervene (for instance, favoring females that could somehow recognize and avoid mating with males carrying the transgene). This approach is very close to a field test.

In addition to this research with fish, similar autocidal methods have been developed and initial tests conducted with fruit flies and mosquitos, including transgenic conditional lethals developed in the yellow fever mosquito—that is, a transgene that, when present in an individual, causes it to die when exposed to certain conditions. A field test was even undertaken recently on Grand Cayman Island in the Caribbean with yellow fever mosquitoes carrying a transgene that kills all offspring as larvae or pupae so that any female that mates with a transgenic male produces no offspring.

There is every reason to think that existing technologies will continue to improve and that occasionally a promising approach—either high-tech or low-tech—will be developed from an entirely new direction; the nature of real innovation is that we cannot predict how it will happen or what it will be, only that it will be different. However, none of these technologies will lessen the flood of biological invasions if they are not combined with better policies to stem their spread and better programs to find new invasions quickly and attempt to eradicate them.

12.6 Will new technologies reverse the trend? A test case

In the 1970s, commercial aquaculturists rearing fish for the market brought two Asian carp species, silver carp and bighead carp, to states along the southern Mississippi River, hoping they would clean their ponds. Flooding soon allowed them to escape from ponds in Arkansas, and they made their way to rivers of the Mississippi Basin. The two species spread northward, raising alarms about their potential impact in the Mississippi and its tributaries. Beginning around 1990, their populations increased enormously in many areas. They are now so numerous that an entrepreneur in Illinois has started (with $2 million in federal assistance) a multimillion-dollar business—the Big River Fish Company—catching Asian carp and shipping them to China, where they are prized as food. The contract with a Chinese importer calls for shipment of 30 million pounds of carp. However, both carp are large filter feeders (often well over 30 pounds). Because juveniles of all fish also feed on the same plankton species that these invaders are consuming, the likelihood of competition with native fish is substantial as the populations grow. Adults of some native species, including one that is endangered, also filter feed on plankton and could be threatened. Research on bighead carp in backwater lakes of the Mississippi and its tributary, the Illinois River, already shows a decline in some bait, commercial, and game fish species. Silver carp also began attracting attention because they jump high into the air in great numbers when they hear motors, occasionally injuring fishermen.

Soon an even bigger threat arose. The Illinois River connects to the Des Plaines River, which in turn connects to Lake Michigan through the 28-mile long Chicago Sanitary and Ship Canal, completed in 1900. This is the only route by which ships can reach the Great Lakes from the Mississippi River and its tributaries. Thus, bighead and silver carp could apparently reach the Great Lakes, where it is feared they would wreak havoc with the native fishes, especially if their populations achieved densities remotely like those in parts of the Mississippi Basin. However, two electric barriers were installed in the canal, designed to kill organisms such as carp by electric shock.

Ecologist Christopher Jerde and his colleagues published research in 2010 in which they sought environmental DNA (eDNA) of bighead and silver carp in 2000 water samples taken from the canal. Their results showed that the electric barriers, although they may have reduced numbers of fish passing through the canal, failed to prevent either species from reaching Lake Michigan. Before this eDNA research, standard capture surveys by electrofishing by management agencies had put the limits of both species at 10 to 15 miles south of the barriers. The eDNA of both carp species was found north of the barriers, within 10 miles of Lake Michigan. Worse, on June 22, 2010, a commercial fisherman caught an adult bighead carp only 8 miles from Lake Michigan, slightly upstream from where eDNA had been found eight months earlier.

These events galvanized environmentalists, who noted the evidence already published showing impacts of these carp on native fishes in the Mississippi drainage. Aside from environmental concerns, stakeholders worried about the welfare of the $7 billion Great Lakes sport fishing industry have been vocal advocates of closing the locks that allow ships (and fish) to pass through the canal into Lake Michigan. The electric barriers are problematic even aside from clearly not being foolproof at blocking Asian carp. Electrical outages are always a worry, and the barriers must occasionally be shut down for maintenance.

After the Jerde team published its results, several groups called for closing the locks in the canal. The State of Michigan, leading this effort, requested that the Supreme Court reopen a 1922 case challenging the right of the State of Illinois to connect the Mississippi system to the Great Lakes in the first place and also requesting a temporary closure of the locks until the court decided whether to reopen the earlier case. The court denied both requests from Michigan. Other motions to close the locks have been denied, and a lawsuit is pending in the U.S. Northern District Court of Illinois demanding immediate action to prevent these carp from entering the Great Lakes. The states of Michigan, Minnesota, Ohio, Pennsylvania, and Wisconsin wanted the U.S. Army Corps of Engineers at least to install nets in two Chicago-area rivers and to expedite research on permanent means of preventing these carp from reaching the Great Lakes, while the lawsuit is pending, but the Supreme Court also denied this request. Meanwhile, in July 2012 President Barack Obama signed legislation ordering the Corps to come up with a plan by January 2014 for blocking the Asian carp's path into the Great Lakes through Chicago waterways and other possible entry points. And in September 2012, the United States and Canada signed an accord to protect the Great Lakes, which sets goals for each country to address invasive species, including Asian carp. Each nation is to implement this agreement through regulations and laws.

The pressure not to close the locks is also great, as both the City of Chicago and Chicago businesspeople and groups have pointed to the economic importance of keeping the shipping channel open between the Mississippi and the Great Lakes. About 12% of Chicago waterways cargo moves by this route. Opponents of closure have taken to impugning the validity of eDNA evidence, demanding "show us the carp." The capture of an adult bighead above the barrier has not quieted that demand. Yet the weight of the scientific evidence is overwhelmingly on the side of the eDNA. It is technically possible that carp DNA could get to an area other than in a living fish. For instance, sewage effluent from humans or excrement from birds that had eaten carp, jettisoning of carp carcasses directly into the water, or release of ballast water containing eDNA that had been picked up below the electric barriers could all carry carp eDNA. However, Jerde and his colleagues show that the exact locations and timing of their finding of eDNA make all these alternatives highly unlikely.

Indeed, the Asian carp case has become a test case for whether political and economic considerations will prevent a powerful new technology from stopping a potentially disastrous invasion. In fact, it is a test case for public response to biological invasions generally. The many technological developments in detecting and managing biological invasions as well as improvements in old methods show that many invasions can be stopped. The success of stringent regulations and rapid response mechanisms, such as those in New Zealand, demonstrate that the flood of introductions can be reduced to a trickle at a cost that is bearable. Economic analyses show that the long-term cost of invasions is staggering and far outweighs the costs of proposed measures to reduce them. The question is whether the public will demand action and whether politicians will be willing to act for the benefit of both society and the environment. Asian carp can be kept out of the Great Lakes. Will this case turn out to be a triumph of foresight and science, or will it be another of the many invasions on which we will look back decades from now and wonder why we let it happen?

APPENDIX

SCIENTIFIC NAMES OF SPECIES CITED IN THIS BOOK

In certain instances, a group of species all in one genus is cited; in such cases, the genus name alone is given.

acacia	*Acacia*	59, 136, 204
African clawed frog	*Xenopus laevis*	76–77
African crystalline ice plant	*Mesembryanthemum crystallinum*	62
African malaria mosquito	*Anopheles gambiae*	105, 187, 196
African rock python	*Python sebae*	133, 165
African wildcat ⸱	*Felis libyca*	83
Africanized honeybee	*Apis mellifera scutellata*	249
Aldabra giant tortoise	*Aldabrachelys gigantea*	234
alder	*Alnus*	72
alewife	*Alosa pseudoharengus*	37, 53, 63, 86–87
alfalfa weevil	*Hypera postica*	117
alligatorweed	*Alternanthera philoxeroides*	214, 215
alligatorweed flea beetle	*Agasicles hygrophila*	214, 215
American bison	*Bison bison*	5, 244

rabbit (Europe)	*Oryctolagus cuniculus*	7, 12, 34, 48, 63, 64, 68, 80, 87, 88, 110, 114–116, 130, 163, 191, 192, 193, 211, 212, 215
rabbit (Americas)	*Sylvilagus*	66
rabbitfish	*Siganus luridus, Siganus rivulatus*	56
raccoon	*Procyon lotor*	65, 143
raccoon dog	*Nyctereutes procyonoides*	131
ragweed	*Ambrosia artemisiifolia*	33
ragwort	*Senecio jacobaea*	126
rainbow trout	*Oncorhynchus mykiss*	13, 16, 38, 77, 82, 131, 143, 147, 234
rat kangaroo	*Bettongia gaimardi, Bettongia penicillata, Bettongia lesueur, Aepyprymnus rufescens*	63
red alga	*Caulacanthus ustulatus*	142
red ant (Great Britain)	*Myrmica sabuleti*	88
redbay ambrosia beetle	*Xyleborus glabratus*	70
redbay laurel	*Persea borbonia*	70
red-bellied blacksnake	*Pseudechis porphyriacus*	67
red-cockaded woodpecker	*Picoides borealis*	46–47, 229
red-crested cardinal	*Paroaria coronata*	237
red deer	*Cervus elaphus*	35, 60, 82–83, 193
red-eared slider	*Trachemys scripta elegans*	132, 169
red fox	*Vulpes vulpes*	62, 66, 163, 211, 212

GLOSSARY

allelopathy: production by a plant of a chemical that affects survival, growth, or reproduction of other individuals of the same or different species

archeophyte: plant species introduced in ancient times, usually considered to be before 1492

ballast: heavy material (usually water or soil) used to improve balance and navigability of ships

biological control: introduction of a natural enemy to control a biological invader; enemy typically imported from native range of the target species

biotype: group of individuals that differ genetically from others of their species in some recognizable feature, such as ability to use a different host plant or to tolerate a pesticide

bottleneck: phenomenon in which a small group of individuals becomes reproductively isolated from others of the same species, as when a few founders constitute an introduced population

coevolution: process in which evolutionary changes in one species trigger evolutionary changes in a second species, which in turn trigger evolutionary changes in the first species, ad infinitum

cryptogenic: of uncertain origin; may have arrived with human assistance or on its own

ecosystem: community of living organisms together with their physical environment

ecosystem engineer: species that changes the physical structure of the environment

endemic: native and exclusive to a particular location

eradication: complete removal of all individuals of a distinct population, not contiguous with other populations

extirpation: elimination of a local population, but with individuals of the same species remaining in contiguous populations or nearby

genetic drift: changes in frequency of a gene variant caused by random processes, such as which copy of a gene in a parent's genome is included in an egg or sperm

genome: all of an individual's hereditary information (genes)

genotype: the genetic makeup of an individual, usually with respect to a specific trait

germplasm: living tissue from which new individuals can be grown, especially seeds

hybridization: mating by individuals of different groups, especially different species but sometimes different populations of the same species

inbreeding depression: reduced fitness in a population caused by mating among related individuals

introduced species: species introduced to a new location with direct or indirect human assistance

invasive species: introduced species that has spread well beyond its arrival point and that perpetuates itself without human assistance

life history trait: traits related to growth and reproduction in a population, such as age at first reproduction or number of offspring

mycorrhizae: fungi that form mutualistic associations with plants after colonizing plant roots

naturalized: term used to describe a population that was introduced long ago and perpetuates itself without human assistance, though not necessarily in pristine "natural" environments

neophyte: plant species introduced relatively recently, often referring to after 1492

pheromone: a chemical emitted by an individual that triggers a response by members of the same species

phytoplankton: planktonic plants (see *plankton*)

plankton: organisms that live and drift in the water, unable to swim

propagule pressure: rate at which individuals of a nonnative species arrive at a site, in terms of numbers of individuals in each arriving group and number of groups

subspecies: distinct groups of individuals of the same species, but usually not breeding with one another, especially by virtue of geographic isolation

transgene: gene or genetic material that has been transferred from one individual to another individual, usually of a different species

trophic cascade: phenomenon in which species A eats species B, thereby reducing the population size of B and lessening its feeding on population C; can even extend to a fourth species if increase of population C leads to increased feeding on population D, which then declines

water molds: group of fungus-like organisms, also known as oomycetes

xenophobia: fear or hatred of foreigners or strangers

zooplankton: planktonic animals (see *plankton*)

NOTES

Chapter 1

1   J. Wolschke-Bulmahn, "Review of R.E. Grese, 'Jens Jensen: Maker of natural parks and gardens.'" *Journal of Garden History* 15 (1995): 54–55.

Chapter 2

1   C. Darwin, *The Voyage of the Beagle* (1845; New York: Bantam Books, Edition 1958), 127.

2   C. K. Yoon, "Alien species threaten Hawaii's environment," *New York Times*, December 29, 1992.

3   F.W. Preston, "On modeling islands," *Ecology* 49 (1968): 592–594, 592.

Chapter 11

1   P. J. Pauly, "The beauty and menace of the Japanese cherry trees," *Isis* 87 (1996): 54.

2   Ibid., 70.

3   K. Ellis, "Save the Australian Pines," (1999) http://www.kodachrome.org/pines.

4   C.S. Sargent, "Editorial," *Garden and Forest* 1 (1888): 266

5   A. Leopold, *A Sand County Almanac, with Essays on Conservation from Round River* (1949; New York: Ballantine Books, edition 1970), 193.

6   Quoted in G. Vince, "Embracing invasives," *Science* 331 (2011): 1383.

7   Animal People Online, "Channel Islands National Park ex-chief hits cruelty of killing 'invasive species,'" (2005) http://www.animalpeoplenews.org/05/4/tsg.channelIslands4.05.htm.

8   F. Messina, "Laguna Niguel trees felled, tempers rise," *Los Angeles Times*, April 14, 1994.

SUGGESTIONS FOR FURTHER READING

1. **General Information**

Cadotte, Marc W., Sean M. McMahon, and Tadashi Fukami, eds. 2006. *Conceptual Ecology and Invasion Biology: Reciprocal Approaches to Nature.* Dordrecht, The Netherlands: Springer.

Dolin, Eric J. 2003. *Snakehead: A Fish out of Water.* Washington, DC: Smithsonian.

Elton, Charles E. 1958. *The Ecology of Invasions by Animals and Plants.* London: Methuen. (reprinted edition 2000. Chicago: University of Chicago Press.)

Richardson, David M., ed. 2011. *Fifty Years of Invasion Ecology.* Oxford: Wiley-Blackwell.

Simberloff, Daniel, and Marcel Rejmánek, eds. 2011. *Encyclopedia of Biological Invasions.* Berkeley: University of California Press.

Spear, Robert J. 2005. *The Great Gypsy Moth War.* Amherst: University of Massachusetts Press.

2. **Magnitude, Geography, and Time Course of Invasions**

Crosby, Alfred. 1986. *Ecological Imperialism: The Biological Expansion of Europe, 900–1900.* Cambridge, UK: Cambridge University Press.

3. **Ecological Effects of Introduced Species—Straightforward Impacts**

Davis, Mark A. 2009. *Invasion Biology.* New York: Oxford University Press.

Francis, Robert A. 2012. *A Handbook of Global Freshwater Invasive Species*. London: Earthscan.

Galil, Bella S., Paul F. Clark, and James T. Carlton, eds. 2011. *In the Wrong Place—Alien Marine Crustaceans: Distribution, Ecology and Impacts*. Dordrecht, The Netherlands: Springer.

Gherardi, F., ed. 2007. *Biological Invaders in Inland Waters: Profiles, Distribution and Threats*. Dordrecht, The Netherlands: Springer.

Hendrix, Paul F., ed. 2006. *Biological Invasions Belowground: Earthworms as Invasive Species*. Dordrecht, The Netherlands: Springer.

King, Carolyn. 1984. *Immigrant Killers. Introduced Predators and the Conservation of Birds in New Zealand*. Auckland: Oxford University Press.

Langor, David W., and Jon Sweeney, eds. 2009. *Ecological Impacts of Non-Native Invertebrates and Fungi on Terrestrial Ecosystems*. Dordrecht, The Netherlands: Springer.

Lockwood, Julie M., Martha F. Hoopes, and Michael P. Marchetti. 2007. *Invasion Biology*. Malden, MA: Blackwell.

4. Impacts of Invasions—Complications and Human Impacts

Keller, Reuben P., David M. Lodge, Mark A. Lewis, and Jason F. Shogren, eds. 2009. *Bioeconomics of Invasive Species*. New York: Oxford University Press.

Perrings, Charles, Harold Mooney, and Mark Williamson, eds. 2010. *Bioinvasions and Globalization*. New York: Oxford University Press.

Pimentel, David, ed. 2011. *Biological Invasions. Economic and Environmental Costs of Alien Plant, Animal, and Microbe Species*, 2nd ed. Boca Raton, FL: Taylor & Francis.

5. Evolution of Introduced and Native Species

Cox, George W. 2004. *Alien Species and Evolution*. Washington, DC: Island Press.

6. How and Why Do Invasions Occur?

Blackburn, Tim M., Julie L. Lockwood, and Phillip Cassey. 2009. *Avian Invasions: The Ecology and Evolution of Exotic Birds*. New York: Oxford University Press.

Gollasch, Stephan, Bella S. Galil, and Andrew N. Cohen, eds. 2006. *Bridging Divides: Maritime Canals as Invasion Corridors*. Dordrecht, The Netherlands: Springer.

Lever, Christopher. 1992. *They Dined on Eland: The Story of Acclimatisation Societies*. London: Quiller.

Ruiz, Gregory M., and James T. Carlton, eds. 2003. *Invasive Species: Vectors and Management Strategies*. Washington, DC: Island Press.

7. Can We Predict Species Invasions?

Franklin, J. 2010. *Mapping Species Distributions: Spatial Inference and Prediction*. Cambridge, UK: Cambridge University Press.

Groves, Richard H., F. Dane Panetta, and John G. Virtue, eds. 2001. *Weed Risk Assessment*. Collingwood, Australia: CSIRO.

Mooney, Harold A., Richard N. Mack, Jeffrey A. McNeely, Laurie E. Neville, Peter Johan Schei, and Jeffrey K. Waage, eds. 2005. *Alien Invasive Species: A New Synthesis*. Washington, DC: Island Press.

National Research Council (USA). 2001. *Predicting Invasions of Nonindgenous Plants and Plant Pests*. Washington, DC: National Academy Press.

8. How Are Species Introductions Regulated?

Bio Intelligence Service. 2011. *A Comparative Assessment of Existing Policies on Invasive Species in the EU Member States and in Selected OECD Countries*. Paris: Bio Intelligence Service.

Miller, Marc L., and Robert N. Fabian, eds. 2004. *Harmful Invasive Species: Legal Responses*. Washington, DC: Environmental Law Institute.

9. Detection and Eradication of Introduced Species

Baskin, Yvonne. 2002. *A Plague of Rats and Rubbervines*. Washington, DC: Island Press.

Buhs, Joshua B. 2004. *The Fire Ant Wars*. Chicago: University of Chicago Press.

Dahlsten, Donald L., and Richard Garcia, eds. 1989. *Eradication of Exotic Pests*. New Haven, CT: Yale University Press.

Meinesz, Alexandre. 2001. *Killer Algae: The True Tale of a Biological Invasion* (with a new postscript). Chicago: University of Chicago Press.

Stolzenburg, William. 2011. *Rat Island.* New York: Bloomsbury.

Veitch, C. Richard, Mick N. Clout, and David R. Towns, eds. 2011. *Island Invasives: Eradication and Management.* Gland, Switzerland: IUCN.

10. Maintenance Management of Invasions

Clout, Mick N., and Peter A. Williams, eds. 2009. *Invasive Species Management.* New York: Oxford University Press.

Dyck, V. Arnold, Jorge Hendrichs, and Alan S. Robinson, eds. 2005. *Sterile Insect Technique.* Dordrecht, The Netherlands: Springer.

Inderjit, ed. 2009. *Management of Invasive Weeds.* Dordrecht, The Netherlands: Springer.

Van Driesche, Roy, Mark Hoddle, and Ted Center. 2008. *Control of Pests and Weeds by Natural Enemies.* Malden, MA: Blackwell.

Wittenberg, Rüdiger, and Matthew J. W. Cock, eds. 2001. *Invasive Alien Species: A Toolkit of Best Prevention and Management Practices.* Wallingford, UK: CAB International.

11. Controversies Surrounding Biological Invasions

Coates, Peter. 2006. *Strangers on the Land: American Perceptions of Immigrant and Invasive Species.* Berkeley: University of California Press.

Gobster, Paul H., and R. Bruce Hull, eds. 2000. *Restoring Nature.* Washington, DC: Island Press.

12. Prospect—The Homogeocene?

Mooney, Harold A., and Richard J. Hobbs, eds. 2000. *Invasive Species in a Changing World.* Washington, DC: Island Press.

Journals

Aliens

Aquatic Invasions

BioControl

Biological Control

Biological Invasions
Diversity and Distributions
Emerging Infectious Diseases
Invasive Plant Science and Management
Journal of Aquatic Plant Management
Natural Areas Journal
Neobiota

Websites

http://www.invasivespeciesinfo.gov	US Department of Agriculture
http://www.invasivespecies.gov	US National Invasive Species Council
http://www.fws.gov/invasives/	US Fish and Wildlife Service
http://www.epa.gov/glnpo/ invasive/	US Environmental Protection Agency
http://www.fs.fed.us/pnw/inva sives/index.shtml	USDA Forest Service Pacific Northwest Research Station
http://www.invasive.org/	University of Georgia Center for Invasive Species and Ecosystem Health
http://plants.ifas.ufl.edu	University of Florida Center for Aquatic and Invasive Plants
http://www.invasivespecies.gc.ca	Government of Canada
http://www.invasivespecies.Centre. ca	Invasive Species Centre (Canada)
http://www.naisin.org	North American Invasive Species Network
http://www.nonnativespecies.org	Great Britain Non-Native Species Secretariat
http://www.europe-aliens.org	Delivering Alien Invasive Species Inventories for Europe

http://ec.europa.eu/environment/ nature/invasivealien/index_en.htm	European Commission
http://www.nobanis.org	European Network on Invasive Species
http://www.biosecurity.govt.nz	Ministry for Primary Industries (New Zealand)
http://www.doc.govt.nz/ conservation/threats-and-impacts/	Department of Conservation (New Zealand)
http://www.environment.gov.au/ biodiversity/invasive/	Government of Australia
http://www.invasives.org.au/	Invasive Species Council (Australia)
http://invasives.org.za	Invasive Species South Africa
http://www.issg.org	Invasive Species Specialist Group (IUCN)

. . .

INDEX

Printed in the USA/Agawam, MA
December 5, 2013